Optical Astronomical Spectroscopy

Series in Astronomy and Astrophysics

Series Editors: **M Birkinshaw**, University of Bristol, UK
M Elvis, Harvard–Smithsonian Center for Astrophysics, USA
J Silk, University of Oxford, UK

The Series in Astronomy and Astrophysics includes books on all aspects of theoretical and experimental astronomy and astrophysics. Books in the series range in level from textbooks and handbooks to more advanced expositions of current research.

Other books in the series

An Introduction to the Science of Cosmology
D J Raine and E G Thomas

The Origin and Evolution of the Solar System
M M Woolfson

The Physics of the Interstellar Medium
J E Dyson and D A Williams

Dust and Chemistry in Astronomy
T J Millar and D A Williams (eds)

Observational Astrophysics
R E White (ed)

Stellar Astrophysics
R J Tayler (ed)

Forthcoming titles

The Physics of Interstellar Dust
E Krügel

Dust in the Galactic Environment, 2nd edition
D C B Whittet

Very High Energy Gamma Ray Astronomy
T Weekes

Dark Sky, Dark Matter
P Wesson and J Overduin

Series in Astronomy and Astrophysics

Optical Astronomical Spectroscopy

C R Kitchin

University of Hertfordshire

CRC Press
Taylor & Francis Group
Boca Raton London New York

CRC Press is an imprint of the
Taylor & Francis Group, an **informa** business

Published in 1995 by
Taylor & Francis Group
711 Third Avenue,
New York,
NY 10017, USA

Published in Great Britain by
Taylor & Francis Group
2 Park Square
Milton Park, Abingdon
Oxon OX14 4RN

First issued in hardback 2017

ISBN 13: 978-1-138-40629-2 (hbk)
ISBN 13: 978-0-7503-0346-0 (pbk)

Library of Congress Cataloging-in-Publication Data

Catalog record is available from the Library of Congress

Taylor & Francis Group
is the Academic Division of Informa plc.

Visit the Taylor & Francis Web site at
http://www.taylorandfrancis.com

and the CRC Press Web site at
http://www.crcpress.com

For Christine with love

Contents

Preface

Despite recent developments in x-ray, gamma ray and long wave astronomy, our picture of the universe is still largely based upon observations made in the optical part of the spectrum (i.e. roughly from 100 nm to 10 μm). Moreover our detailed understanding of the physics of the processes occurring on and within stars, planets, nebulae, galaxies, etc. comes largely from the use of the spectroscope to study the details of the lines in those objects' spectra. It is no exaggeration to say that three-quarters or more of astronomical knowledge would be unknown if the optical spectroscope had never been invented. Yet a perusal of most general astronomical texts would lead to the conclusion that spectroscopy hardly existed, filled as they are with direct images. I hope that this book will go some way towards setting that record straight.

The book is intended to provide a grounding in all aspects of the theory and practice of modern astronomical spectroscopy over the optical part of the spectrum. Additionally, reference is made to aspects of spectroscopy outside this range at times for completeness (e.g. molecular spectra) or because of importance (e.g. the hydrogen 21 cm line). It is anticipated that the book will be useful to those already engaged in astronomical spectroscopy as a reference work, to intending professional astronomers as a textbook to introduce them to the subject, and to amateur astronomers as a challenge to extend their work out of the usual direct observations. Sufficient physical background is included throughout to enable the astronomical aspects of the topic to be understood, but it is not intended to give a thorough coverage of the physical and chemical aspects of spectroscopy.

The book is not aimed at a specific level, though a good knowledge of astronomy, at least up to that of thorough familiarity with the material in one of the many introductory astronomy books intended for first year university courses, is assumed. In addition some knowledge of mathematics and physics is required.

I hope you, the reader, find the book useful and interesting, and I wish you well in your future use of this, the most fundamental tool of astrophysics.

C R Kitchin
March 1995

Part 1

Atomic Processes

1

Introduction to Spectroscopy

1.1 HISTORICAL BACKGROUND

Most people's first, and many people's only, experience of a spectrum occurs with the sight of a rainbow. The school child's mnemonic: 'Richard Of York Gave Battle In Vain' for the traditional colours of the rainbow (Red–Orange–Yellow–Green–Blue–Indigo–Violet) also reminds us that in the popular imagination there are only seven colours to the spectrum. The rainbow was eventually correctly explained in 1304 by Theodoric of Freiberg, though there had been earlier attempts by Aristotle and Roger Bacon. Theodoric showed by experimenting with water in a spherical flask that the primary bow originates when water droplets in the atmosphere internally reflect incoming sunlight once (figure 1.1), the different colours, as we now know, arising from the change in refractive index of water with wavelength. The second bow arises from two internal reflections. Descartes and Newton provided complete explanations in terms of geometrical optics for the primary and secondary bows. The supernumerary arcs (the pink and green stripes sometimes seen just inside the primary bow), however, are an interference phenomenon and have only recently been satisfactorily explained. Less commonly observed, though at least as frequent as rainbows, when one knows what to look for, are the sundogs or parhelia. Sundogs appear as rather poor spectra at an angle of around 25° to the Sun, and arise from ice crystals rather than droplets.

The phenomenon of differential refraction, which underlies rainbows and sundogs, became of more than aesthetic significance with the invention of the refracting telescope. The simple lenses used in early telescopes produced a series of coloured images along the optical axis, rather than a single sharply focused white light image. This effect is known as chromatic aberration, and it much reduced the usefulness of those early telescopes. In 1666 Newton showed that a beam of sunlight passed through a triangular glass prism produced a purer spectrum than the rainbow. Moreover he also went on to show that a second reversed prism recombined the colours back to white light, proving that white light was composite. Regrettably, he did not use a fine enough beam of light to

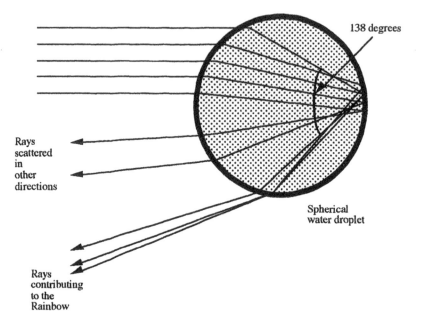

Figure 1.1 Production of a rainbow. The rays are actually sent out in many directions. The visible rainbow arises because of the concentration of rays around the direction near 138° to the incoming light.

observe the dark lines present in the solar spectrum, or quantitative spectroscopy might have started a century and a half earlier than it did.

After Newton, there was a long hiatus in significant advances in spectroscopy until the nineteenth century, though Thomas Melville did discover the emission spectra of flames in the early eighteenth century. Then, in 1800, William Herschel discovered the infrared part of the spectrum by showing that a thermometer still registered energy when moved beyond the red end of the solar spectrum. Similarly the ultraviolet was discovered a year later by Johann Ritter when he observed that the breakdown of silver nitrate into silver, caused by light, occurred even more rapidly when that substance was placed beyond the violet end of the solar spectrum.

The dark lines missed by Newton were eventually found in 1802 by William Wollaston, but he mistakenly took them to be the natural boundaries between the colours.

We should note at this point that the spectral *line* is not a phenomenon of the spectrum itself. The dark portions of the spectrum arise from the absence or reduction of energy at a particular wavelength. The shape of that dark portion in the image of the spectrum then just mirrors the shape of the entrance aperture. Since most spectroscopes use slits as their entrance apertures, we get lines in the resulting spectrum. However, if other shapes of aperture were used, the

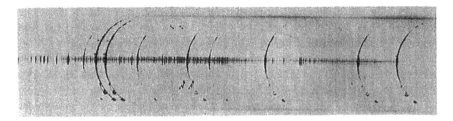

Figure 1.2 Flash spectrum of the solar chromosphere taken with a slit-less spectroscope (reproduced by permission of the Royal Astronomical Society).

Figure 1.3 Fraunhofer's solar spectrum (taken from *A Short History of Astronomy* by A Berry, figure 97 (Dover, 1961)).

Table 1.1 The Fraunhofer lines.

Line	Wavelength (nm)	Origin
A	759.4	Molecular oxygen (terrestrial)
B	686.7	Molecular oxygen (terrestrial)
C	656.3	Hydrogen (Hα)
D1	589.6	Sodium
D2	589.0	Sodium
E1	527.0	Iron
Eb	518.3–516.7	Magnesium triplet
F	486.1	Hydrogen (Hβ)
G	430.8	Iron plus CH molecule (now, G′ is often used for Hγ at 434.0)
H	396.8	Calcium
K	393.3	Calcium
L	382.0	Iron ⎱
M	373.5	Iron ⎰ Added after Fraunhofer's time
N	358.1	Iron ⎰

spectral features would also take on other shapes. Thus for example in the flash spectrum of the solar chromosphere, the 'lines' (in this case emission lines) appear as arcs (figure 1.2).

Quantitative spectroscopy thus really began with Joseph von Fraunhofer. In 1814 he used a slit and a theodolite to produce a beam of sunlight, which he then allowed to fall onto a prism. He saw nearly 600 lines in the solar spectrum. His discoveries were not solely due to his use of a slit, but also because he had developed vastly improved ways of casting glass. His prisms were therefore better than any previously available. By 1823 he was able to measure wavelengths and he mapped 324 of the solar lines (figure 1.3). He also labelled the nine most prominent lines, or groups of lines, with the capital letters of the alphabet, and those lines are still to this day known as the Fraunhofer lines. The sodium D lines and the calcium H and K lines are the best known examples (table 1.1). Fraunhofer came within a hair of anticipating Kirchhoff and Huggins' work when he observed dark lines in the spectra of stars. However, he found the lines were different from those to be found in the Sun, and was unable to explain them.

The solar spectrum was first photographed in 1842 by Alexandre Becquerel, and a decade later Jean Foucault demonstrated that light shone through a sodium flame contained dark lines coincident with the D lines in the solar spectrum. But the next major development in understanding took a surprisingly long time from Fraunhofer's original discoveries. It was not until 1859 that Gustav Kirchhoff, working with Robert BunsenRobert, published the fundamental law of quantitative spectroscopy which we now know as Kirchhoff's law:

The ratio between the powers of emission and the powers of absorption for rays of the same wavelength is constant for all bodies at the same temperature.

In modern mathematical terms,

$$\frac{\varepsilon_\lambda(T)}{k_\lambda(T)} = \text{constant} \tag{1.1}$$

where $\varepsilon_\lambda(T)$ is the emission coefficient, and $k_\lambda(T)$ is the absorption coefficient of the substance at wavelength λ and temperature T. To the fundamental law, we have the corollaries:

The wavelengths emitted by a substance depend upon that substance and the temperature.

and

The absorption of a substance is a maximum at those wavelengths which it tends to emit.

and

A luminous solid or liquid (or a compressed gas) emits a continuous spectrum, whereas a gas gives a spectrum consisting of bright lines.

With this new understanding, quantitative spectroscopy advanced rapidly.

Within a couple of years the new elements, caesium and rubidium, were discovered on the Earth by Kirchhoff and Bunsen from their spectra. Kirchhoff also inferred the existence of sodium within the Sun from the presence of the D lines in its spectrum, and later found there another half dozen elements, including iron. In 1862 Anders Ångstrom identified hydrogen in the Sun. Then in 1864 William Huggins identified the presence of hydrogen, iron, sodium and calcium in stars from the lines in their spectra. Thus the fiat pronounced by Auguste Comte in 1825, that the chemical composition of the stars was an indisputable example of the kind of knowledge that mankind would never possess, was triumphantly refuted by the first major discovery of astronomical spectroscopy.

After Kirchhoff's breakthrough, insights into all branches of spectroscopy came thick and fast. Huggins, a self-taught amateur astronomer, made a remarkable series of discoveries by applying spectroscopy to astronomy. In the same year, 1864, that he found the composition of stars, he also showed that a planetary nebula was composed of hot glowing gas by observing its emission line spectrum. The bright comet of that year also had an emission spectrum, thus proving it too to contain a thin gas. His subsequent observations of some 70 nebulae over the next two years showed emission line spectra in about a third of the cases, and stellar type spectra—dark lines on a bright background— in the rest. Thus was the problem of the nature of those nebulae that could not be resolved into stars finally solved. Observations of subsequent comets enabled Huggins to identify the presence in them of compounds of hydrogen and carbon. He also observed that the emission spectrum was superimposed on a faint background solar spectrum, showing that the comet reflected light as well as emitting it.

Huggins had to develop and build most of his own equipment for these observations. One of his innovations, which almost every spectrograph from that day to this still incorporates, was the ability to produce a comparison spectrum. This is a second spectrum produced by an artificial light source (an electric spark in Huggins' case), the wavelengths of whose lines are precisely known, and from which the wavelengths of the star's lines may be found by comparison. By being able to measure the wavelengths of the lines in a spectrum to a high degree of precision, Huggins was able to solve yet another intractable problem, that of how to determine the component of the velocity of an astronomical object along the line of sight. His determination of velocity was via the shifts from their normal positions of spectral lines arising from the Doppler effect as the object moves towards or away from us. In 1868 he measured the line-of-sight velocity of Sirius as $47 \, \text{km s}^{-1}$. Huggins also pioneered the use of photography in spectroscopy, developed spectroscopic methods of observing the solar chromosphere, prominences and corona, observed the spectrum of a nova, and investigated the near ultraviolet spectra of stars in a lifetime's work that entitles him to be considered the father of astronomical spectroscopy.

Huggins, however, was not alone in his pioneering work. Herman Vogel also measured stellar radial velocities, and in 1871 measured the Doppler shifts at the approaching and receding limbs of the Sun due to its rotation—a remarkable achievement since the wavelengths of the lines change by only 0.003 nm. In 1889 Antonia Maury discovered that Mizar (ξUMa) was a binary star from the periodic doubling of the lines in its spectrum as one star approached and the other receded from us in their mutual orbital motion. Soon afterwards Vogel showed that Algol (β Per) was also a binary, as had long been expected from its brightness variations, by observing the cyclical wavelength shifts of the lines in its spectrum. For Algol, only one star's lines were visible, and so there was no line doubling as in the case of Mizar. Another significant advance was the production of high quality gratings by Henry Rowland from 1882 onwards, resulting in improved spectroscopes.

Norman Lockyer postulated in 1869 the existence of an as then unknown element, from the presence of its lines in the solar spectrum. The new element was named helium, and it was not found on Earth for another three decades. The example of helium led to the inference of other new elements: Coronium from unknown lines in the solar corona, and Nebulium from lines in the spectra of interstellar nebulae. However, these new elements did not exist, and the reason for the unknown lines was only elucidated this century. The lines were due to atoms in very high levels of ionization (Coronium) or to forbidden transitions (chapter 4) in atoms of existing elements (Nebulium).

In 1863 a different approach to astronomical spectroscopy, whose descendants continue to be used to this day, was initiated by Angelo Secchi. This was the classification of stars according to the appearance of their spectra. Secchi's classification was simple: Type I, spectra with only the lines of hydrogen visible (generally white stars); Type II, spectra similar to that of the Sun (generally yellow stars); Type III, spectra with bands shaded towards the red (generally red stars); and Type IV, spectra with bands shaded towards the violet. That classification was soon extended, and stars became sorted according to the complexity (number of lines) in their spectra. Class A were thus stars with the least number of lines in their spectra, Class B the stars with slightly more lines and so on. By the end of the nineteenth century Edward Pickering was engaged in cataloguing stellar spectra in enormous numbers using objective prism spectra. This work eventually resulted in the Henry Draper Catalogue, which with its extension contains information on a third of a million stellar spectra.

The classification used in the HD catalogue forms the basis of the spectral classification system that is still in use today. Unfortunately, the previous system based upon spectral complexity was adapted, rather than a new system being started from scratch. Since we now classify stellar spectra on the basis of the temperature of the stars producing those spectra, the classification classes have been shuffled around, some have been eliminated and others added, resulting in the present system being very difficult to use. The order of spectral classes (chapter 11) now in use is thus (going from high to low temperature):

$$O - B - A - F - G - K - M$$

The order can be remembered from the mnemonic **Oh Be A Fine Girl/Guy Kiss Me!**, but it would have been preferable for a more logical system to have been chosen in the first place.

In another hangover from the early days of astronomical spectroscopy, in which the sequence was thought to be an age sequence as well, stars of classes O, B and A are often referred to as Early Stars, and those of classes K and M as Late Stars. While a convenient usage, it should be stressed that this terminology does not now have any evolutionary implications.

These astronomical discoveries were being paralleled by advances in understanding the nature of atoms and hence of the physical processes underlying the production of spectra. Thus in 1885 Johann Balmer found an empirical law to predict the wavelengths of the hydrogen lines in the visible spectrum (now known as the Balmer lines):

$$\lambda = 364.56 \frac{n_2^2}{n_2^2 - n_1^2} \text{ (nm)} \tag{1.2}$$

where n_1 and n_2 are small integers. In its more familiar modern version

$$\nu = R\left(\frac{1}{n_1^2} - \frac{1}{n_2^2}\right) \tag{1.3}$$

where R is the Rydberg constant and has a value of 3.288×10^{15} Hz, n_1 has values of 1, 2, 3, 4 etc corresponding to the Lyman, Balmer, Paschen, Brackett etc series for hydrogen and n_2 takes integer values from $n_1 + 1$ upwards, the equation correctly predicts the other series of hydrogen lines outside the visible spectrum.

Joseph Thomson's discovery of the electron in 1897, Max Planck's explanation of the black-body energy distribution by requiring radiation to be in the form of packets of energy (quanta) in 1900, Albert Einstein's confirmation of the existence of quanta through his explanation of the photoelectric effect five years later, and Ernest Rutherford's investigations of the structure of atoms in 1908 culminated in 1913 in Niels Bohr's model of the atom. The restriction of the allowed energies (orbits) for electrons in atoms in this model explains why an atom absorbs or emits only at specific wavelengths. The observed wavelengths then correspond to the energy gaps between pairs of permitted energies for the electrons. The relationship between energy and wavelength had been found earlier by Planck:

$$e = h\nu = \frac{hc}{\lambda} \tag{1.4}$$

where h is Planck's constant (6.626×10^{-34} J s) and c is the velocity of light (2.998×10^8 m s^{-1}).

The Bohr model of the atom, though inadequate to explain all spectroscopic phenomena, is clearly recognizable as the forerunner of today's ideas.

Those modern ideas are covered in later parts of this book, and therefore will not be considered in any depth here. The main developments, however, include Sommerfeld's extension of Bohr's model, quantum mechanics, Pauli's exclusion principle, Heisenberg's uncertainty principle, understanding the effects of ionization, explaining molecular spectra and the explanation of the Zeeman effect.

While these developments in the observation and understanding of the optical (i.e. near infrared, visual and near ultraviolet) spectrum were occurring, elsewhere the concept of the spectrum was being extended very much more widely. Thus in 1860 James Clerk Maxwell produced his theory predicting the properties of radiation associated with electrical and electromagnetic phenomena. In this theory, light becomes only a small component of the full spectrum, which has wavelengths ranging down towards zero (x- and gamma rays) and outwards towards infinity (microwaves and radio waves). The radio waves were discovered experimentally in 1887 by Heinrich Hertz, and x-rays in 1895 by Wilhelm Röntgen.

The nature of the interaction of electromagnetic radiation with matter varies with wavelength, even though there is no change in the fundamental nature of the radiation. Thus, simplistically, in the optical region we are concerned primarily with electronic transitions in atoms and molecules. At longer wavelengths we have direct induction of currents in conductors, and the rotational and vibrational transitions in molecules, while at short wavelengths the interactions are with the inner electrons of atoms and with their nuclei. These interactions are much less accessible than those in the optical region. The development of spectroscopy outside the optical region was therefore quite slow. Thus it was not until 1933 that any significant work was carried out even in the laboratory on the microwave spectroscopy of gases. At short wavelengths, detectors such as Geiger counters and nuclear emulsion were available much earlier, but had poor energy discrimination. Even today, spectroscopy of high energy gamma rays has spectral resolutions of 10% or worse.

The astronomical applications of spectroscopy outside the optical region were even slower than those in the laboratory. Karl Jansky detected the first radio waves from an astronomical object when in 1933 he detected the galactic centre. However, it was not until after the second world war that radio astronomy really got under way. Microwave, ultraviolet, x- and gamma ray work had to wait until the 1960s, when rockets and spacecraft could lift the detectors above the Earth's atmosphere, before any significant astronomical progress could be made.

Thus astronomical spectroscopy varies from a well established technique in the optical region (though not without its problems still), through a recently developed but quite well understood application in the radio region, to a still developing technique at the shortest wavelengths.

1.2   TYPES OF SPECTROSCOPY

The fundamental spectroscopic techniques of interest to astronomers are atomic and molecular absorption and emission spectroscopy. Fluorescence spectroscopy can also crop up very occasionally, and the spectral properties of solids are of interest within the solar system and for the study of the interstellar medium. The principles of these areas of spectroscopy are briefly summarized below, and dealt with in detail later in this chapter and elsewhere within this book. The other areas of spectroscopy have not currently found any astronomical applications. They are summarized briefly below both for completeness and because references to them are likely to be encountered when reading spectroscopic texts from non-astronomical sources.

1.2.1 Atomic absorption spectroscopy (AAS)

This is the basic form of spectroscopy employed in astronomy. It is dealt with later in detail. In summary, an atom or ion absorbs radiation at specific frequencies whose energies (equation 1.4) are given by the differences between the quantized energy levels of the electrons in the atom, as those electrons jump from lower to higher energies.

Emission spectroscopy studies the emission lines produced as electrons jump downwards in atoms and ions. In molecules, as well as the emission or absorption arising from electron transitions, radiation can be produced or absorbed as the vibration and/or rotation of the molecule changes between quantized values.

1.2.2 Continuum processes

Numerous processes produce continuous or pseudo-continuous emission or absorption of radiation. Ionization of atoms and ions, for example, causes broad absorption bands in stellar spectra. The formation of the H^- ion (i.e. the 'recombination' of a neutral hydrogen atom with a second electron to form a negative ion) is of considerable personal interest because it provides the bulk of the visible continuum emission from the Sun. The acceleration of charged particles, usually electrons spiralling around magnetic fields, produces the continuum emission, or more rarely absorption, known as synchrotron radiation. The same process operating in the vicinity of the electric fields of atoms and ions produces free–free or bremsstrahlung radiation. Charged particles moving faster than the local speed of light in a material produce Čerenkov radiation. Hot solids, liquids and dense gases emit radiation in a pattern that is often close to black-body (equation (4.7)) spectra (but see 'solid state spectroscopy' (1.2.10) below). The Compton and inverse Compton effects result in the transfer of energy back and forth between photons and electrons and can 'smear-out' line spectra (as in the solar coronal spectrum). Lastly, but by no means finally, the vibrational and rotational absorptions and emissions (chapter 5) of molecules can sometimes give the appearance of continuous bands when observed using a spectroscope with too low a resolution or dispersion to separate the individual lines.

1.2.3 Nuclear magnetic resonance (NMR) spectroscopy

NMR spectroscopy studies the behaviour of those atomic nuclei which have magnetic moments arising from their possessing a net spin (chapter 2). These are nuclei with odd atomic masses, or even mass number and odd atomic charge, e.g. ^1H, ^{13}C, ^{14}N, ^{31}P etc. In the presence of an external magnetic field, quantized orientations of the nucleus to the magnetic field have differing energies. Transitions between these states can occur through the absorption or emission of radiation just as for electrons producing lines in 'normal' spectra. The energies involved for the magnetic field strengths generally used (a tesla or so), are such that the transitions are in the radio region, with frequencies of a few tens of MHz. In the laboratory, NMR spectroscopy uses a strong magnetic field and a radio frequency transmitter to illuminate the sample. Either the magnetic field strength or the transmitter frequency is varied, and the resonant frequency detected by a radio receiver. The closely related technique of nuclear quadrupole resonance spectroscopy (NQR) is based upon the quantization of the nuclear electric quadrupole moment with respect to electric charge. The electric field in NQR spectroscopy is derived from the nearby atoms, and the observed frequencies range from 100 kHz to 1 GHz. NQR requires that the nucleus has a quadrupole moment, which implies that its nuclear spin quantum number be greater than 1/2. Thus it is applicable to nuclei such as ^{14}N, ^{33}S, ^{35}Cl etc.

1.2.4 Raman spectroscopy

Raman spectroscopy is based upon inelastic scattering by molecules. An intense monochromatic beam of light which does *not* coincide with one of the molecule's lines, usually nowadays from a laser, is used to illuminate the sample. The radiation scattered sideways is then observed. The scattered radiation will contain a strong component at the original frequency, due to Rayleigh scattering, but also much weaker components at frequencies a little lower and higher than the original, due to Raman scattering. These secondary components are known as the Stokes lines when they are of lower frequencies, and anti-Stokes lines when they are of higher frequencies than that of the exciting line. The Stokes lines arise from energy being subtracted from the illuminating radiation as the molecule undergoes an upward vibrational transition (chapter 5) during the scattering process. The anti-Stokes lines arise when energy is added to the illuminating radiation by downward transitions. The difference between the frequency of the exciting radiation and those of the Stokes or anti-Stokes lines is thus just the frequency of the vibrational transitions involved. Rotational transitions (chapter 5) introduce fine structure to the lines. The technique thus allows transitions that would normally produce lines in the infrared or microwave part of the spectrum to be observed in the visible. It also provides information on transitions that are difficult or impossible to excite in normal absorption or emission spectroscopy.

1.2.5   Electron spin resonance spectroscopy (ESR)

ESR spectroscopy, like NMR (see above), is based upon transitions between quantized orientations to an external magnetic field. ESR spectroscopy, as its name implies, however, utilizes electrons rather than nuclei, and is confined to the study of atoms, ions and molecules having one or more unpaired electrons. The experimental technique is similar to that of NMR spectroscopy, except that with the typically used magnetic field strengths of a few tenths of a tesla, the resonant frequencies are a few tens of GHz.

1.2.6   Phosphorescence spectroscopy

In phosphorescence, an excited molecule decays (undergoes a series of downward transitions) towards the ground state, emitting radiation from various transitions as it does so.   However, the decay path includes a metastable (chapter 4) state. The lifetime (chapter 4) of that state is in the region of a few seconds (compared with 10^{-8} seconds or so for 'normal' transitions). Thus some of the emission occurs with a significant delay when compared with other parts of the emission sequence. The phenomenon is principally used to study the energy levels of triplet states within molecules. The technique involved uses a light source with an appropriate wavelength to excite the molecule. That light source is chopped, and the emission from the sample detected with an appropriate delay at the same chopping frequency, in order to isolate the phosphorescent emissions.

1.2.7   Fluorescence spectroscopy

The phenomenon of fluorescence is closely related to phosphorescence (see above). An excited atom or molecule decays towards the ground state, emitting radiation from the various transitions as it does so.   The difference from phosphorescence is that no metastable state is involved, and so the emissions all occur within 10^{-8} seconds or so of the original excitation. The downward transitions can be at the same frequency as the exciting photons or at intermediate frequencies (figure 1.4).

The mineral Fluorspar, which emits in the visible when illuminated by ultraviolet radiation, is a well known example of fluorescence. In the laboratory, the phenomenon allows the intermediate energy levels between the ground state and the excited state to be studied.

Fluorescence occurs occasionally in astronomical situations, when it is often known as selective or resonant fluorescence. The effect occurs when an emission line from one element coincides, or nearly coincides, with the wavelength of a transition of another element.   The second element then absorbs from the emission line and the upper level becomes overpopulated compared with the normal equilibrium situation. Downward transitions from that upper level then result in emission lines appearing in the spectrum of the object at intermediate

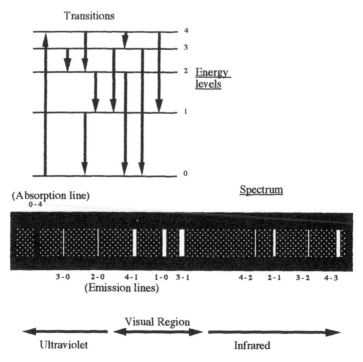

Figure 1.4 A schematic diagram showing the transitions between different energy levels (a Grotrian diagram, chapter 3), resulting in fluorescence. The spectrum contains one exciting line in the ultraviolet, and two fluorescing lines in the ultraviolet, three in the visible and four in the infrared.

wavelengths. An example of this occurs in the B type emission line stars (Be stars). Strong emission in the hydrogen Lyman-β line at 102.572 nm is absorbed by a transition of neutral oxygen at 102.577 nm because the line-widths are sufficiently broad for the two lines to overlap almost completely. Downward routes back to the ground state involve emissions at wavelengths of 102.5, 115.7, 130.2, 483.4, 844.6 and 1128.7 nm.

1.2.8 Mössbauer spectroscopy

This technique is also known as nuclear gamma ray resonance (NGR) spectroscopy, and involves transitions between excited states *within a nucleus*. The energies involved are thus far larger than those so far considered, and the radiation that is involved is in the gamma ray region. Transitions between quantized excited states (and the ground state) can occur within a nucleus just as with electrons within an atom, and in a similar manner lead to the absorption and emission of energy. However, the energies of gamma rays, and hence their momenta, are sufficiently high that the nucleus recoils significantly during such a transition. The absorption or emission thus does not occur at a specific frequency, but over a range of frequencies determined by the recoil velocity. Mössbauer discovered in 1958 that under certain conditions the recoil of a single nucleus could be shared amongst all the atoms of a crystal containing that nucleus. The recoil thus effectively reduces to zero, and the emitted or absorbed frequency becomes extremely sharp. The observational technique requires the detector (a crystal containing absorbing nuclei) be moved towards or away from the source until the Doppler shift causes the emitted and absorbed frequencies to coincide. The radiation is then strongly absorbed, and the velocity at which this matching or resonance occurs gives the wavelength shift between the emitted radiation and the absorbing frequency of the nuclei.

1.2.9   Photoelectron spectroscopy

In the photoelectric effect, photons of a sufficiently high energy are absorbed by a solid, and electrons emitted in their stead. The energy of the photons has to be above the minimum required to excite the electrons and enable them to escape from the surface of the solid (this minimum energy is known as the work function). The energy of the escaping electrons is then the difference between the photon's energy and the work function. If monochromatic radiation of high enough energy is used to illuminate the sample, then the emitted electrons will have energies corresponding to the differences between the photon energy and the work functions for each of the valence electrons. Vibrational transitions will impose a fine structure upon the emitted energies. Photons in the x-ray region enable the inner electron energy levels to be studied in a similar manner. The electron energies are analysed by using electrostatic fields to focus electrons of a specific energy onto a detector. The field strength is then varied so that electrons of other energies are brought to the detector. Thus the whole electron energy spectrum may be built up, and from that the atom or molecule's energy levels inferred.

1.2.10   Spectroscopy of solids

This topic is also considered in more detail in chapter 7. Here, we just note that solids and liquids do have features in their absorption spectra. Those features, however, are usually very broad and variable. They do not, in general, provide unique signatures identifying particular elements and compounds. This is, of course, a matter of everyday experience—we are aware of the spectra of solids through their colours, and that these vary with conditions (lighting, surface texture, thickness, humidity etc).We are also familiar with our inability to distinguish between some materials just on the basis of their appearance ('Not all … that glisters, (is) gold', etc! (quoted from 'Ode on the Death of a Favourite Cat' by T Gray)).

2

The Physics of Atoms and Molecules

The structures of atoms and molecules are fundamental to the formation of their spectra. Those structures are determined by the principles of quantum mechanics, and in particular by the quantization of the electron's energy levels. Quantization requires that an electron in an atom can have only certain specific energies, not the infinitely variable values allowed by classical physics.

For most purposes in astronomical spectroscopy, the Bohr–Sommerfeld model of the atom is adequate. Since this is conceptually much easier than the full-blown quantum mechanical description of an atom, we shall employ the Bohr–Sommerfeld model whenever possible. It gives accurate results for hydrogen and hydrogen-like ions, but quantum mechanics is needed to describe more complex atoms.

On the Bohr–Sommerfeld model, the electrons are envisaged as orbiting the small, massive, positively charged nucleus, like the planets orbit the Sun (figure 2.1). Electrons in different orbits have different energies; quantization therefore corresponds to the electrons being permitted only certain orbits. In quantum mechanical terms we can understand quantization as permitting only those orbits within which an integer number of the de Broglie wavelengths (equation (2.23)) for the electron can be fitted (figure 2.2). In such orbits a standing wave will be set up. In other orbits, there would be destructive interference between different parts of the wave, and it would be destroyed.

In allowed orbits, Bohr postulated that the electron's angular momentum (mvr) had to be an integral multiple of $h/2\pi$ (where m is the electron's mass $(9.110 \times 10^{-31}$ kg), v its velocity, and r the orbital radius for a circular orbit). Thus

$$mvr = n\frac{h}{2\pi}. \tag{2.1}$$

The electron is attracted to the nucleus by the Coulomb force, and in a circular orbit experiences a constant acceleration, so that

$$\frac{Ze^2}{4\pi\varepsilon_0 r^2} = \frac{mv^2}{r} \tag{2.2}$$

14

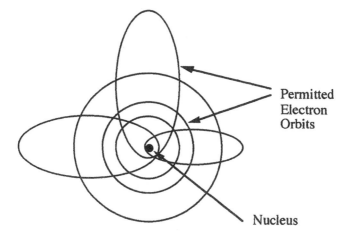

Figure 2.1   The Bohr–Sommerfeld model of the atom.

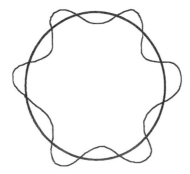

Figure 2.2   The standing de Broglie wave of an electron in a permitted orbit.

where e is the charge on the electron (1.602×10^{-19} C), ε_0 the permittivity of free space (8.854×10^{-12} F m^{-1}), and Z the atomic number of the nucleus. Thus

$$r = \frac{Ze^2}{4\pi \varepsilon_0 m v^2} \tag{2.3}$$

and so from equation (2.1), eliminating v,

$$r = \frac{\varepsilon_0 n^2 h^2}{\pi Z e^2 m}. \tag{2.4}$$

Thus, the physical size of the orbits is proportional to n^2, and inversely proportional to the charge on the nucleus. For hydrogen ($Z = 1$), the first few circular orbits have the radii given in table 2.1. The number n is called the Principal Quantum Number.

Table 2.1 Radii of the lower Bohr orbits in hydrogen.

n	r (nm)
1	0.053
2	0.212
3	0.477
4	0.848
etc	

Table 2.2 Energies of the lower Bohr orbits in hydrogen.

n	E $(10^{-18}$ J$)$	E (eV)
1	−2.18	−13.60
2	−0.54	−3.40
3	−0.24	−1.51
4	−0.14	−0.85
etc		

The electron's energy in such an orbit may be found by summing its potential and kinetic energies. By convention, however, the potential energy is regarded as zero when the electron is in the outermost orbit ($r = \infty$). In practice, this means when the atom is ionized, and so has lost the electron concerned. Since potential energy is released as the electron gets nearer to the nucleus, the potential energies of electrons in actually allowed orbits are negative and given by

$$\text{Potential energy} = -\frac{Ze^2}{4\pi\varepsilon_0 r} \tag{2.5}$$

Thus from equation (2.2), we find

$$E = -\frac{Ze^2}{4\pi\varepsilon_0 r} + \frac{Ze^2}{8\pi\varepsilon_0 r} = -\frac{Ze^2}{8\pi\varepsilon_0 r} \tag{2.6}$$

and so from equation (2.4)

$$E = -\frac{Z^2 e^4 m}{8\varepsilon_0^2 n^2 h^2}. \tag{2.7}$$

The commonly used unit for electron energies is the electron-volt (eV), which is the energy gained by an electron when accelerated by a potential difference of one volt. One eV is equal to 1.602×10^{-19} J. The energies of the first few circular orbits in hydrogen in both J and eV are given in table 2.2.

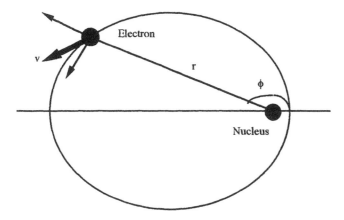

Figure 2.3 The elliptical orbit of an electron.

From equation (2.7), we may see that for a given atom, such as hydrogen, the energy depends only on n^{-2}. The differences in energies between two levels are therefore proportional to $(n_1^{-2} - n_2^{-2})$, where n_1 and n_2 are the quantum numbers (integers) for the two levels. Thus we get the Rydberg formula (equation (1.3)) for the wavelengths of the hydrogen lines.

2.1 SOMMERFELD'S REFINEMENTS

The simple treatment above needs three refinements to enable it to fit the known data reasonably precisely.

Firstly, the electron does not orbit about the centre of the nucleus, but about the centre of mass of the system. This effect may be allowed for by replacing the mass of the electron in equations (2.1) to (2.7) by the 'reduced mass', μ:

$$\mu = \frac{mM}{m + M} \qquad (2.8)$$

where M is the mass of the nucleus. Thus for hydrogen, for which $M = 1.672 \times 10^{-27}$ kg, $\mu/m = 0.99945$, and the correction is only in the fourth significant figure.

More significantly, Sommerfeld postulated the existence of elliptical orbits within which the integral of the momentum of the electron around the orbit was an integer multiple of h. In an elliptical orbit, an electron has two degrees of freedom, and each of these is separately quantized. In polar coordinates, the electron's position may be specified by its radial distance from the nucleus, r, and its azimuthal angle, ϕ (figure 2.3). The potential energy is again given by equation (2.5), but the kinetic energy is now given by

$$\text{Kinetic energy} = \frac{mv^2}{2} = \frac{m}{2}(\dot{r}^2 + r^2 \dot{\phi}^2). \qquad (2.9)$$

The momentum now has two components:

$$p_\phi = mr^2\dot\phi \qquad (2.10)$$

for the azimuthal component, and

$$p_r = m\dot r \qquad (2.11)$$

for the radial component. Applying Sommerfeld's postulate, the integrals of each of these momenta around the orbit must multiples of Planck's constant;

$$\int_{\phi=0}^{2\pi} p_\phi \, d\phi = kh \qquad (2.12)$$

$$\int_{\phi=0}^{2\pi} p_r \, dr = zh \qquad (2.13)$$

where k and z are integers and are called the Azimuthal and Radial Quantum Numbers, respectively.

Now, for motion under an inverse square attraction, the angular momentum is constant (cf Kepler's second law of planetary motion), and so equation (2.12) gives us

$$2\pi p_\phi = kh = 2\pi mr^2\dot\phi = 2\pi mvr \qquad (2.14)$$

and this is identical to Bohr's postulate for circular orbits (equation 2.1), with k replacing n. Evaluating equation (2.13) is a lengthier process. But we may eventually find

$$\frac{1}{\sqrt{1-\varepsilon^2}} - 1 = \frac{z}{k} \qquad (2.15)$$

where ε is eccentricity of the ellipse ($\varepsilon^2 = (a^2 - b^2)/a^2$ where a and b are the semi-major and semi-minor axes of the ellipse respectively), which gives

$$1 - \varepsilon^2 = \frac{k^2}{(k+z)^2} = \frac{b^2}{a^2}. \qquad (2.16)$$

Thus the eccentricity of the ellipse is quantized, since k and z may take only integer values.

When $z = 0$, we have $\varepsilon = 0$, and the ellipse reduces to Bohr's circular orbit. A value of k of zero would reduce the orbit to a straight line. Since this would require the electron to pass through the nucleus, such a value is impossible, and actual values of k start from 1. As we shall see, the size of the orbit (i.e. its semi-major axis) is now proportional to $(k+z)^2$, and by analogy with equation (2.4) we may identify $(k+z)$ with the principal quantum number, n, defined previously. Thus equation (2.16) gives

$$\frac{b}{a} = \frac{k}{n} \qquad (2.17)$$

and we find that the allowed elliptical orbits are just those whose major and minor axes are in the ratio of simple integers.

The two quantum numbers for the Bohr–Sommerfeld model now become the principal quantum number, $n(= 1, 2, 3, 4, 5, \ldots)$ and the azimuthal quantum number, $k(= 1, 2, 3, \ldots n)$. The energy of an electron in such an orbit is given to a first approximation by the same equation as for Bohr's circular orbits (equation (2.7)), so that the energy does not depend on the azimuthal quantum number.

However, the third refinement to the basic Bohr model is to allow for relativistic effects, which cause the orbits to precess. Taking relativity into account, there then is a weak dependence of the energy upon k. This results in a splitting of any resulting spectral lines, which is known as their fine structure. The relativistic effect is thus dependent upon a constant known as the fine structure constant, α:

$$\alpha = \frac{e^2}{2\varepsilon_0 hc} = 7.2974 \times 10^{-3}. \tag{2.18}$$

The electron's energy is then given by

$$E = -\frac{Z^2 e^4 \mu}{8\varepsilon_0^2 n^2 h^2}\left[1 + \frac{\alpha^2 Z^2}{n}\left(\frac{1}{k} - \frac{3}{4n}\right)\right] \tag{2.19}$$

where the reduced mass, μ, has been used in place of m in equation (2.7).

The change in energy is small: the energies for the first excited level of hydrogen ($n = 2$) differ between the circular orbit ($k = 2$) and the only allowed elliptical orbit ($k = 1$) by only 4.5×10^{-5} eV, corresponding to a wavelength difference of about 0.01 nm for lines in the visible part of the spectrum.

The length of the semi-major axis is the same as the radius of the circular orbit (equation (2.4) with μ in place of m). Denoting the radius of the first circular orbit for hydrogen by a_0, we have

$$a_0 = \frac{\varepsilon_0 h^2}{\pi e^2 \mu} = 5.294\,688 \times 10^{-11} \text{ (m)} \tag{2.20}$$

and so

$$a = a_0 \frac{n^2}{Z} \tag{2.21}$$

and

$$b = a_0 \frac{kn}{Z}. \tag{2.22}$$

Thus the first few orbits for the hydrogen atom have the shapes and sizes shown in figure 2.4 .

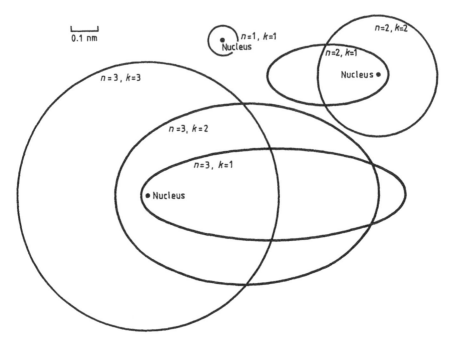

Figure 2.4 The shapes and sizes of the Bohr–Sommerfeld orbits for hydrogen for $n = 1$, $k = 1$; $n = 2$, $k = 1, 2$; and $n = 3$, $k = 1, 2, 3$. For other atoms the sizes would be reduced inversely with Z.

2.2 SPACE QUANTIZATION

Two further quantum numbers are required to specify the energy of an electron within an atom completely. The first of these arises because the atom is three-dimensional. This third quantum number determines the orientation of the electron's orbit in space. In the absence of any reference direction, all the quantized orientations will have the same energy (i.e. the energy levels will be degenerate). However, when there is a reference direction, such as when a magnetic field is present, some or all of the orientations will have different energies. This phenomenon results in the Zeeman and Stark effects (chapter 6).

Quantum mechanics obtains a quantum number analogous to the azimuthal quantum number, k, of the Bohr–Sommerfeld model (see below). This quantum number, l, is also confusingly called the azimuthal quantum number (or sometimes the *reduced* azimuthal quantum number), and is equal to $(k - 1)$. The quantum mechanical azimuthal quantum number therefore runs from 0 to $(n - 1)$.

We may specify the orientation of the orbit by the direction of a vector normal to the orbital plane (for further justification of this, see chapter 3). Anticipating the quantum mechanical treatment (see below), the quantization of

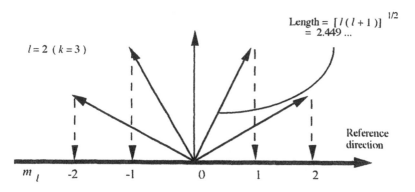

Figure 2.5 Space quantization of the orientation of an electron's orbit. The projections of the vector perpendicular to the orbital plane onto the reference direction must be integer multiples of $h/2\pi$.

the orientation of the orbit is then such that the projection of that vector's length of $[l(l+1)]^{1/2}$ onto the reference direction must be an integer (figure 2.5). The allowed values of the quantum number deriving from the orientation of the orbit, m_l, thus run $0, \pm 1, \pm 2, \ldots \pm l$, or from 0 to $\pm(k-1)$. This quantum number, because the energy levels become non-degenerate most commonly in a magnetic field, is called the Magnetic Quantum Number. The orbit vector precesses about the reference direction at the Larmor frequency, $\nu_L (= 1.400 \times 10^{10} H$ Hz, where H is the magnetic field strength in tesla). So there is no quantization of the other angular coordinate because the vector takes up all positions.

2.3 ATOMIC STRUCTURE

The structures of the atoms comprising the elements are determined by the limitations on allowed electron orbits, together with one other principle. This latter is the Pauli Exclusion Principle. It states that

No two electrons in an atom can possess exactly the same energy and spin.

The principle is an empirical one, justified by the correctness of the predictions that result if it is assumed to be true.

Now as we have seen, the three quantum numbers, n, l (or k) and m_l, determine an electron's energy. The electron also behaves as though it is spinning, and that spin is quantized at values of $\pm h/4\pi$. Spin quantization has been confirmed by the decisive Stern–Gerlach experiment. In this experiment, a beam of alkali metal atoms was split into two beams by the action of a magnetic field on the two possible values of the electron's spin. The spin of the electron thus becomes the fourth quantum number, m_s, and takes values of $\pm 1/2$ (since its magnitude of $(3^{1/2}h/4\pi)$ is half the 'normal' quantization unit). There is a small dependence of the electron's energy upon m_s, as discussed further below (see section on Multiple Electron Atoms). The Pauli Exclusion Principle becomes with the introduction of the spin quantum number,

No two electrons in an atom may have the same four quantum numbers, n, l (or k), m_l, and m_s.

In this last form, the exclusion principle determines the properties of all the known elements, and underlies their arrangement into the periodic table. The chemical properties of the elements are governed by the outermost electrons, and the exclusion principle requires that in most cases the distribution of those outermost electrons changes from element to element.

The energy of an orbit is primarily determined by the principal quantum number, n (equation (2.19)). The other quantum numbers introduce minor variations. Thus we may most usefully consider orbits in increasing order of the value of n. As a reminder, given the value of n, the other quantum numbers can then only take values:

$$l = 0, 1, 2, 3, \ldots, (n-1) \qquad \{k = 1, 2, 3, 4, \ldots, n\}$$
$$m_l = 0, \pm 1, \pm 2, \pm 3, \ldots, \pm l$$
$$m_s = \pm 1/2.$$

For $n = 1$, we must therefore have $l = 0$, $m_l = 0$ and $m_s = \pm 1/2$. A maximum of two electrons can thus be fitted into the lowest energy orbit, and they must have opposite spins. Thus we get hydrogen with one electron, and helium with two electrons, both in the $n = 1$ orbit.

For $n = 2$, we may have $l = 0$ or 1. From equation (2.19) (remembering $k = l + 1$, and that the energy is negative), we may see that the electron's energy decreases as l decreases. The next lowest energy orbit after $n = 1$, $l = 0$, therefore has $n = 2$, $l = 0$. The magnetic quantum number is again limited to 0, and only two electrons may be put into this orbit and they must have opposing spins. Thus we get lithium and beryllium, each with two electrons in the lowest energy orbit, and with one and two electrons in the next orbit respectively. For $n = 2$, $l = 1$, m_l may take values 0 and ± 1. With m_s taking values $\pm 1/2$, we may therefore put up to six electrons into this next orbit. Thus the elements boron, carbon, nitrogen, oxygen, fluorine and neon are formed with the next one to six electrons.

For $n = 3$, l may take the values 0, 1 or 2 in order of increasing energy. For $l = 0$ or 1, the orbits may again be filled with up to two or six electrons respectively. Thus the elements sodium and magnesium, and then aluminium to argon are produced. For $l = 2$, m_l may take the values 0, ± 1 or ± 2, and m_s again can be $\pm 1/2$. Thus up to ten electrons can go into this orbit.

The next element, potassium, however, is not formed by an electron going into the $n = 3$, $l = 2$ orbit. Instead, the electron goes into the $n = 4$, $l = 0$ orbit, because the latter has a lower energy than the former (figure 2.6). Calcium is likewise formed with two electrons in the $n = 4$, $l = 0$ orbit with opposite spins. The other elements follow in a similar fashion. As may be seen from figure 2.6, the next orbit after $n = 4$, $l = 1$, will be $n = 3$, $l = 2$, and the ten electrons which can be fitted in there produce the elements scandium to zinc, the first set of transition elements. Thereafter we have

$n = 4$, $l = 1$ orbit (gallium to krypton),

$n = 5$, $l = 0$ (rubidium and strontium),

$n = 4$, $l = 2$ (the second set of transition elements, yttrium to cadmium),

$n = 5$, $l = 1$ (indium to xenon),

$n = 6$, $l = 0$ (caesium and barium),

$n = 4$, $l = 3$ (the rare earths, cerium to ytterbium),

$n = 5$, $l = 2$ (lutetium, and the third set of transition elements, hafnium to mercury),

$n = 6$, $l = 1$ (thallium to radon),

$n = 7$, $l = 0$ (francium and radium),

$n = 5$, $l = 3$ (the actinides, thorium to lawrencium).

Thus the broad features of the periodic table are determined. Within this overall pattern there are further anomalies, such as lanthanum with an electron in the $n = 5$, $l = 2$ orbit, which appears between barium and cerium, but such finer details need not concern us here.

2.4   OTHER VIEWPOINTS

The study of atomic structure was not undertaken solely by atomic physicists, but also by chemists and spectroscopists. Since the three areas were developed independently of each other, the representations of atomic structure and its consequences were also developed independently. The formalism of atomic physics, based upon quantum numbers, is the most fundamental, and is now widely used in all areas of study. However, other notations will be encountered and some are in common use.

Thus in chemistry, the periodic table (see above) was developed initially from the chemical properties of the elements. The columns in that table correspond to different values of the azimuthal quantum number, and the rows to different values of the principal quantum number. Electrons with the same value of n are spoken of as being within the same *shell*. The shells are labelled with capital letters, starting from K. Thus $n = 1$ electrons are in the K shell, $n = 2$ electrons in the L shell, $n = 3$ electrons in the M shell, and so on (figure 2.6).

The chemical properties of elements are primarily determined by the electrons in the outermost shell. Thus the transition elements, the actinides and the rare earths, where electrons are being added to inner shells, leaving the outermost electrons mostly unchanged, are very similar to each other chemically, and many of the latter were only discovered from their spectra.

The study of spectroscopy began by looking at spectra. Some elements, notably the alkalis, produced sets of lines of differing appearance or pattern.

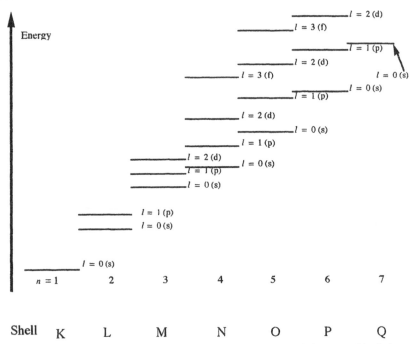

Figure 2.6   Schematic representation of the relative energies of electron orbits in known atoms.

These sets became labelled as Sharp, Principal, Diffuse and Fundamental. Subsequently, it was found that these corresponded to values of l of 0, 1, 2 and 3. Electrons with those values of l consequently have become called s, p, d or f electrons (figure 2.6, see also chapter 3). Electrons with values of l of 4 and upwards are similarly labelled by lower case letters, starting alphabetically from g. The principal quantum number is just given by its numerical value.

Under these alternative notations, we find that the electron structure of iron (for example) may be given as

2 electrons with $n = 1$, $l = 0$
2 electrons with $n = 2$, $l = 0$
6 electrons with $n = 2$, $l = 1$
2 electrons with $n = 3$, $l = 0$
6 electrons with $n = 3$, $l = 1$
2 electrons with $n = 4$, $l = 0$
6 electrons with $n = 3$, $l = 3$

as

2 electrons in the K shell
8 electrons in the L shell
14 electrons in the M shell

2 electrons in the N shell

or as

2× 1s electrons
2× 2s electrons
6× 2p electrons
2× 3s electrons
6× 3p electrons
6× 3d electrons
2× 4s electrons.

2.5   QUANTUM/WAVE MECHANICS

As already mentioned, the Bohr–Sommerfeld model, while it gives a good description of the behaviour of hydrogen and hydrogen-like ions (atoms with all but one of their electrons removed), is not so successful with more complex atoms. Two new theories were put forward in the early 1920s which led to improved descriptions. These theories, Wave Mechanics (due to de Broglie and Schrödinger) and Quantum Mechanics (due to Heisenberg) were later shown to be two different mathematical formulations of the same idea.

De Broglie put forward the idea that *particles* could behave like *waves*. That *waves* could behave like *particles* had been suggested two decades earlier by Planck, and amongst other phenomena is demonstrated by the photoelectric effect. Nowadays, the wave nature of particles is a relatively commonplace idea, and underlies, for example, the operating principles of electron microscopes. The wavelength associated with a particle, the de Broglie wavelength, is given by

$$\lambda = \frac{h}{mv}. \tag{2.23}$$

As we have already seen (figure 2.2), we can then picture the quantization of Bohr–Sommerfeld orbits as being just those orbits with a stable standing de Broglie wave in them.

The quantum numbers come from the solutions to the equation describing the wave properties of the electrons. That equation is the Schrödinger wave equation:

$$\frac{\partial^2 \Psi}{\partial x^2} + \frac{\partial^2 \Psi}{\partial y^2} + \frac{\partial^2 \Psi}{\partial z^2} = \frac{1}{v^2} \frac{\partial^2 \Psi}{\partial t^2} \tag{2.24}$$

where Ψ is the displacement of the wave, v, the wave velocity, x, y and z the position coordinates and t time. Solutions to equation (2.24) may be found when ψ takes the form

$$\Psi = \psi e^{2\pi i v t / \lambda} \tag{2.25}$$

where λ is the de Broglie wavelength (equation (2.23)) and ψ the amplitude of the wave. ψ must everywhere be single valued, continuous, finite and must vanish at infinity for real solutions. In physical terms, the Bohr–Sommerfeld orbits are now replaced by the probability of finding an electron in a given volume, dv, which is given by $\psi\psi^* dv$, where ψ^* is the complex conjugate of ψ.

We then get the time-independent Schrödinger equation

$$\frac{\partial^2\psi}{\partial x^2} + \frac{\partial^2\psi}{\partial y^2} + \frac{\partial^2\psi}{\partial z^2} + \frac{4\pi^2\mu^2 v^2}{h^2}\psi = 0 \tag{2.26}$$

where μ is the reduced mass (equation (2.8)).

In polar coordinates, the equation becomes

$$\frac{1}{r^2}\frac{\partial}{\partial r}\left(r^2\frac{\partial\psi}{\partial r}\right) + \frac{1}{r^2\sin^2\theta}\frac{\partial^2\psi}{\partial\phi^2} + \frac{1}{r^2\sin^2\theta}\frac{\partial}{\partial\theta}\left(\sin\theta\frac{\partial\psi}{\partial\theta}\right) + \frac{4\pi^2\mu^2 v^2}{h^2}\psi = 0 \tag{2.27}$$

and we may find solutions for hydrogen-like ions by taking ψ to be separately dependent on functions of ϕ, θ and r:

$$\psi = \Phi\Theta R. \tag{2.28}$$

equation (2.27) then becomes

$$\frac{\partial}{\partial r}\left(r^2\frac{\partial R}{\partial r}\right)\frac{1}{R} + \frac{1}{\sin^2\theta}\frac{\partial^2\Phi}{\partial\phi^2}\frac{1}{\Phi} + \frac{1}{\sin\theta}\frac{\partial}{\partial\theta}\left(\sin\theta\frac{\partial\Theta}{\partial\theta}\right)\frac{1}{\Theta} + \frac{4\pi^2\mu^2 v^2}{h^2}r^2 = 0. \tag{2.29}$$

The term involving ϕ is independent of ϕ, since by separating it out in equation (2.29), we may see that it depends only on θ and r. We thus get

$$\frac{1}{\Phi}\frac{\partial^2\Phi}{\partial\phi^2} = \text{constant} = C_1. \tag{2.30}$$

Since ϕ is an angle, solutions to equation (2.30) must be periodic with period 2π, and so take the form

$$\Phi = C_2 e^{iq\phi} \tag{2.31}$$

where C_2 is a constant and q an integer.

Thus we find that the only allowed solutions for Φ occur for integer values of q, and that this constant may therefore be identified with what we have previously called the magnetic quantum number (m_l).

We may similarly argue that the terms in equation (2.29) in θ are functions of ϕ and r, and so independent of θ. We may further substitute equation (2.31) in equation (2.30) to find

$$\frac{1}{\Phi}\frac{\partial^2\Phi}{\partial\phi^2} = -m_l^2 \tag{2.32}$$

and so we get

$$\frac{1}{\Theta \sin \theta} \frac{\partial}{\partial \theta} \left(\sin \theta \frac{\partial \Theta}{\partial \theta} \right) - \frac{m_l^2}{\sin^2 \theta} = \text{constant} = C_3. \tag{2.33}$$

This last equation may be shown to have meaningful solutions only when

$$C_3 = -l(l+1) \tag{2.34}$$

where l is an integer such that $l \geqslant |m_l|$. Thus we obtain the azimuthal quantum number from the only allowed solutions of equation (2.33), and we find that for a given value of l, m_l may take the values $0, \pm 1, \pm 2, \ldots \pm l$. We may now see, as previously discussed, that the quantum mechanical azimuthal quantum number, l, is related to the Bohr–Sommerfeld azimuthal quantum number, k, by $l = k - 1$.

Now the total energy of the electron, E, is given by the sum of its kinetic and potential energies, T and V, and the kinetic energy is just

$$T = \frac{1}{2}\mu v^2 \tag{2.35}$$

and so

$$\mu v^2 = 2(E - V) \tag{2.36}$$

Substituting equations (2.34) and (2.36) into equation (2.29) then gives us the third equation, just in r:

$$\frac{1}{r^2} \frac{\partial}{\partial r} \left(r^2 \frac{\partial R}{\partial r} \right) - \frac{l(l+1)}{r^2} R + \frac{8\pi^2 \mu}{h^2} (E - V)R = 0. \tag{2.37}$$

This equation has meaningful solutions for *negative* values of E, *only* for the values

$$E = -\frac{Z^2 e^4 \mu}{8\varepsilon_0^2 n^2 h^2} \tag{2.38}$$

where n is a positive integer.

By comparison with equation (2.7) we may therefore see that n in equation (2.38) is just the previously identified Bohr–Sommerfeld principal quantum number. Furthermore, equation (2.38) has solutions for *all* positive values of E, and these solutions correspond to the loss of the electron to the atom by ionization.

The Schrödinger equation is thus soluble, not for unrestricted values of E, but only for certain specific values. Those values of E for which the equation may be solved are termed the eigenvalues, and the corresponding functions, ψ, the eigenfunctions. The first three quantum numbers arise quite naturally in quantum mechanics as giving rise to values of the electron's energy which permit solutions to the wave equation, instead of being the arbitrary inventions of the Bohr–Sommerfeld theory.

3

Atomic Energy Levels

The energies of electrons in atoms (and ions and molecules) are fundamental to spectroscopy, for it is by changing its energy within an atom that an electron emits or absorbs photons and so produces the emission or absorption lines that we observe in spectra. Unfortunately the determination of those energies from the properties (quantum numbers) of the electrons is a complex task with many special cases and exceptions. Understanding of the process is further hampered by an archaic and illogical system of notation, which has its origins in the observations of the early spectroscopists made before much theoretical understanding had been achieved. Since, however, that system of notation is in universal use, and all tables of data etc make use of it, the student is forced to come to terms with its idiosyncrasies.

The energies that electrons can take within atoms are quantized. That is to say, only certain restricted values are permitted. As we have seen, Bohr and Sommerfeld introduced quantization as an arbitrary invention to explain the observations. Quantum theory however showed that the Bohr–Sommerfeld quantization arose naturally as the allowed solutions (eigenvalues) of the wave equation, when the electron was thought of as behaving as a wave rather than a particle. Under both systems, the electron's properties are described by a set of four integers, known as quantum numbers. Initially, we are going to be concerned with just three of the quantum numbers; those we have called the Principal Quantum Number (n), the Azimuthal Quantum Number (l) and the Spin (m_s).

Generally, the main factor determining the electron's energy is the value of its principal quantum number. Thus, for example, we have seen for hydrogen (equation (2.7) and table 2.2) that the energy for $n = 1$ is -13.60 eV, while for $n = 2$ it is -3.40 eV. Changing the azimuthal quantum number from $l = 0$ to $l = 1$ alters the energy by only 4.5×10^{-5} eV (equation (2.19)—remember that l is actually the reduced azimuthal quantum number, $l = k - 1$).

3.1 MULTIPLE ELECTRON ATOMS

So far we have largely been concerned with hydrogen-like ions, where there is only a single electron in orbit around the nucleus. Many of the results for such ions can also be applied to the valence electron in the alkali metals, which have just a single electron in their outermost orbit. However, there are differences arising from the presence of the inner electrons which cause the outer electron to be moving in an electric field that deviates from that of a bare nucleus (i.e. effectively from that of a point charge). We have seen that for hydrogen there is weak dependence of the electron's energy upon the azimuthal quantum number (equation (2.19)). Representing these different orbits in an energy level diagram (figure 3.1; a Grotrian diagram—see also figure 2.1), we find that the fine structure does not show up on the scale of that diagram. However for the alkali metals, the adjacent levels have significantly different energies (figure 3.2). The different energies for the levels may be thought of as arising from the outer electron in the more elliptical orbits penetrating closer to the nucleus (figure 2.5) through the cloud of inner electrons. The differences between levels reduce sharply as n increases, and this effect may be thought of as due to the physical size of the orbits (equations (2.21) and (2.22)) increasing to the point where the electron becomes sufficiently far from the cloud of inner electrons that it experiences a good approximation to the electric field of a point source again.

As mentioned earlier, there is a small dependence of the electron's energy upon the value of the spin quantum number, m_s. This dependence arises because the spin and orbital motion interact with each other. The total angular momentum quantum number of the electron, j, is given by

$$j = |l \pm m_s|. \tag{3.1}$$

Thus for levels with $l = 0$, j can only have the value $+1/2$. However, for higher values of l, j will take two values, 3/2 and 5/2 for example when $l = 2$. Each of the levels shown in figures 2.7 and 2.8, except for the first columns, is therefore actually double. Two (or more) such related energy levels are called a Term.

The energy difference arising from different values of j is generally very small. For lithium, for example, the $n = 2$, $l = 1$ term has two energy levels of 1.847 84 eV and 1.847 88 eV, resulting in optical spectrum lines separated by only 0.01 nm. In sodium though, the splitting of the corresponding term results in the well known D lines at 589.0 and 589.6 nm.

For hydrogen-like and alkali-metal-like atoms and ions, the formation of the term is straightforward. Only one electron is involved because the inner electrons (if any) form complete shells or subshells and can be ignored (see later discussion of closed shells).

For hydrogen and hydrogen-like ions, which have just one electron, that

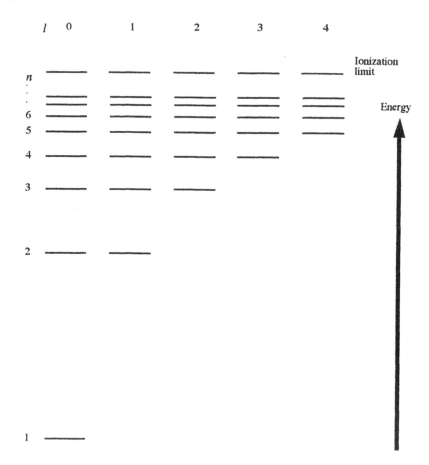

Figure 3.1 Schematic energy level diagram for hydrogen-like ions, showing the (lack of much) effect of the azimuthal quantum number.

electron's quantum numbers determine its energy directly. However, in many-electron atoms, the electrons interact with each other in very complex ways in determining individual energy values. We must therefore consider the atom as a whole in such cases, and obtain quantum numbers appertaining to the atom rather than individual electrons. In considering such interactions, we deal with the orbital angular momentum and the spin angular momentum *vectors* (hereinafter shown in bold: *l* and *s*—remember that angular momentum is given by the vector cross product of velocity and distance from the rotation axis; these angular momentum vectors are therefore perpendicular to the orbital/rotational planes). The actual angular momenta are given by the magnitudes of *l* and *s*:

$$[l(l + 1)]^{1/2} \times (h/2\pi) \qquad [s(s + 1)]^{1/2} \times (h/2\pi). \qquad (3.2)$$

When the electrons in an atom interact with each other, then in the commonest

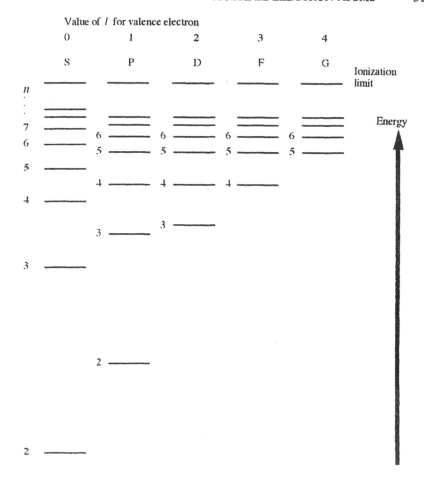

Figure 3.2 Schematic energy level diagram for lithium, showing the effect of the azimuthal quantum number. Similar diagrams would be obtained for other alkali metals and like ions.

case, the individual orbital angular momenta, l_i, combine to produce a resultant orbital angular momentum for the atom, L. The spin angular momenta, s_i, similarly combine to produce a resultant spin angular momentum for the atom, S. These two whole atom angular momenta then combine to produce the total angular momentum of the atom, J. These three angular momenta each have associated quantum numbers, L, S and J, known as the Total Azimuthal Quantum Number, the Total Spin Quantum Number and the Total Angular Momentum Quantum Number (or Inner Quantum Number) respectively. The values of the angular momenta are then given by

$$[L(L+1)]^{1/2} \times (h/2\pi) \tag{3.3}$$

$$[S(S + 1)]^{1/2} \times (h/2\pi) \tag{3.4}$$

and

$$[J(J + 1)]^{1/2} \times (h/2\pi). \tag{3.5}$$

This form of interaction of the electrons is known as Russell–Saunders or L–S coupling. L–S coupling is a good approximation for the lighter elements, and is generally therefore of most interest to astronomers.

At the other extreme, we have j–j coupling. In this situation, the orbital angular momentum, l, and spin angular momentum, s, of each electron combine to produce a total angular momentum for each electron, j (cf equation (3.1)). The j_i then combine to give the total angular momentum of the atom, J. However, the total number of energy values resulting from j–j coupling is the same as if L–S coupling had applied, though the individual energy values differ. This is a result of Ehrenfest's adiabatic law:

For a virtual, infinitely slow alteration of the coupling conditions, the quantum numbers of the system do not change.

In other words, the correct number and type of energy values may be found in all cases by assuming the interactions of individual angular momenta to be zero (or very small). Thus L–S coupling may be assumed in finding the number and type of an electron's energy values for a particular atom.

3.2 L–S COUPLING

Under this form of interaction between the electrons, the angular momentum of each electron interacts with that of every other electron to produce the orbital angular momentum for the atom. Similarly the spin angular momenta of the electrons interact to produce the spin angular momentum of the atom. Clearly, for any but the lightest elements, the number of electrons involved in the calculations is going to become formidable. Fortunately, as we shall justify later, electrons in closed shells and subshells do not contribute to the angular momentum of the atom. Such electrons may be ignored, and only the outermost electrons (valence electrons) normally need to be considered. Thus the problem rarely involves more than four or five electrons in practice.

Let us consider first the orbital angular momenta. The orbital angular momentum for each electron, as we have seen, is given by the vector, l, perpendicular to the Bohr orbital plane, and of magnitude $[l(l+1)]^{1/2} \times h/2\pi$. If we have two electrons, then their individual orbital angular momentum vectors add together by normal vector addition to produce the orbital angular momentum for the atom, L (figure 3.3). The addition is constrained by the requirement that the magnitude of L, $[L(L+1)]^{1/2} \times h/2\pi$, must be such that the total azimuthal quantum number, L, is a positive integer or zero (figure 3.4). Thus in the illustrated case when $l_1 = l_2 = 1$, L can only take the values, 0, 1 or 2. We

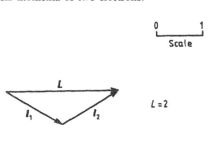

Figure 3.3 Formation of the total orbital angular momentum of the atom, *L*, from the individual orbital angular momenta of two electrons.

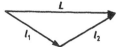

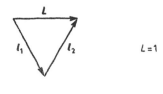

Figure 3.4 Allowed values of the total azimuthal quantum number, *L*, for two electrons with individual azimuthal quantum numbers, $l_1 = l_2 = 1$.

may easily see that for just two electrons the allowed values of *L* will be given by

$$L = (l_1 + l_2), (l_1 + l_2 - 1), (l_1 + l_2 - 2), \ldots, |l_1 - l_2|. \tag{3.6}$$

So, for example, if $l_1 = 2$ and $l_2 = 4$, *L* may take the values 2, 3, 4, 5 or 6 (figure 3.5). In a simpler case, if $l_1 = 0$, then clearly $L = l_2$.

Thus, for a pair of electrons, the resulting orbital angular momentum for the atom may in general take several values, through, in Bohr terms, changing the relative orientations of the electron's orbits, *without the electrons changing their*

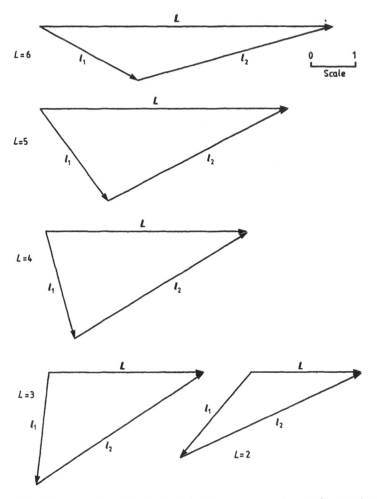

Figure 3.5   Allowed values of the total azimuthal quantum number, L, for two electrons with individual azimuthal quantum numbers, $l_1 = 2$, $l_2 = 4$.

quantum numbers. The different values of L will nonetheless usually correspond to different energies for the electrons in the atom, when considered as a whole.

When more than two electrons are involved, L must be found by combining the orbital angular momentum vectors of two electrons as above, then adding the third to each configuration for the two electrons, and finding all the allowed values for L in each case (figure 3.6). A fourth electron then has its orbital angular momentum vector added to all possible configurations derived for three electrons, and so on. Thus in the illustrated case, with $l_1 = l_2 = 1$, $l_3 = 2$, we have $L = 0, 1, 1, 2, 2, 2, 3, 3$ or 4. L may thus arrive at the same numerical value by several different routes. The energies corresponding to such identical

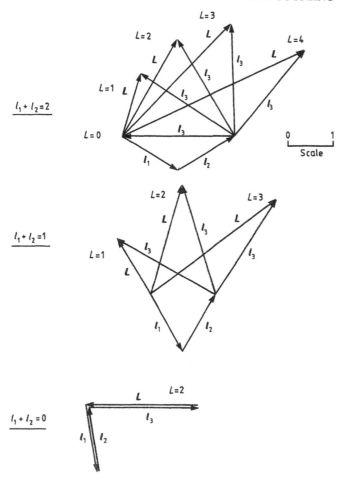

Figure 3.6 Allowed values of the total azimuthal quantum number, L, for two electrons with individual azimuthal quantum numbers, $l_1 = l_2 = 1$, $l_3 = 2$. (NB: the vector diagrams show all possible total orbital angular momenta, *L*, for each allowed combination of l_1 and l_2.)

values of *L*, however, will generally differ from each other.

The spin angular momentum for the atom, *S*, is similarly found from the spin angular momenta of the electrons. These are given by the individual vectors, s_i, of the electrons. However, the situation is slightly simpler, because the values of m_{s_i} are limited to $\pm 1/2$, i.e. $|s_i| = \pm 3^{1/2}(h/4\pi)$. Thus *S* can only take the values

$$S = (1/2 + 1/2 + 1/2 + \ldots),\ (1/2 + 1/2 + 1/2 + \ldots - 1),$$
$$(1/2 + 1/2 + 1/2 + \ldots - 2), \ldots, 0 \text{ or } 1/2 \qquad (3.7)$$

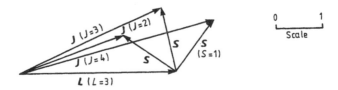

Figure 3.7 Formation of one term of the total angular momentum of an atom with $l_1 = 2$ and $l_2 = 4$.

where each electron not in a closed shell or subshell contributes one of the '1/2s'.

Thus for one electron we must have $S = 1/2$, for two electrons $S = 1$ or 0, for three electrons $S = 3/2$ or $1/2$, etc.

Finally, under L–S coupling we find the total angular momentum of the atom given by the inner quantum number, J, by combining each value of L with each value of S. For a given pair of L and S values, we thus have

$$J = (L + S), (L + S - 1), (L + S - 2), \ldots, |L - S|. \tag{3.8}$$

For $l_1 = 2$ and $l_2 = 4$, then as we have seen, L may take the values 2, 3, 4, 5 or 6. For two electrons, $S = 1$ or 0, giving (figure 3.7)

$$
\begin{aligned}
L &= 2 \quad S = 1 \quad J = 3, 2 \text{ or } 1 \\
L &= 2 \quad S = 0 \quad J = 2 \\
L &= 3 \quad S = 1 \quad J = 4, 3 \text{ or } 2 \\
L &= 3 \quad S = 0 \quad J = 3 \\
L &= 4 \quad S = 1 \quad J = 5, 4 \text{ or } 3 \\
L &= 4 \quad S = 0 \quad J = 4 \\
L &= 5 \quad S = 1 \quad J = 6, 5 \text{ or } 4 \\
L &= 5 \quad S = 0 \quad J = 5 \\
L &= 6 \quad S = 1 \quad J = 7, 6 \text{ or } 5 \\
L &= 6 \quad S = 0 \quad J = 6.
\end{aligned}
$$

The set of energy levels corresponding to the values of J for a given pair of L and S is called a Term.

The two electrons just considered thus give rise to ten terms, half of which contain three values of J, and hence three differing energy values, whilst the others contain just one value of J, and so only one value for the energy. The different mutual interactions of these electrons therefore lead to 20 energy levels, without the value of the principal quantum number (n) of either electron changing. The terms with three energy levels are called triplets, and those with one energy level, singlets. Likewise a term with two energy levels would be a doublet etc. The number of values of J in a term, and hence the number of its energy levels, is generally given by $(2S + 1)$, and this quantity is therefore called the Multiplicity.

In general the resulting energy will be different for every value of J, *even for those numerically identical values of J obtained by different combinations of L and S.*

3.3 SPACE QUANTIZATION

We have seen in the case of a single electron (figure 2.6) that in the presence of a magnetic field the orientation of the orbit is quantized. The quantum number describing the orientation, the magnetic quantum number m_l, takes the values $0, \pm 1, \pm 2, \ldots, \pm l$. The energy of the electron in general is then different for each value or state of m_l. In the absence of a magnetic (or electric) field, the individual states revert to identical energies, and the energy level corresponding to that value of l becomes degenerate with $(2l + 1)$ states within it.

The same phenomenon occurs for multi-electron atoms. This time it is the total angular momentum of the atom whose direction is quantized with reference to a magnetic or electric field. The total angular momentum is given by the inner quantum number vector, J. The orientation of J with respect to the field is then such that the magnitude of its projection onto the field can only take the values $-J(h/2\pi)$, $(-J + 1)(h/2\pi)$, $(-J + 2)(h/2\pi), \ldots, J(h/2\pi)$ (figure 3.8). The quantization of the orientation is described by the fourth quantum number, the magnetic quantum number, M, and this therefore takes the values $-J, -J + 1, -J + 2, ..., J$. Since J can be both integer and half integer, so also can the values of M in the presence of a field. The individual values of M represent different energy values, giving rise to the Zeeman and Stark (chapter 6) effects. The quantization of the orientation of the total angular momentum is independent of the field strength, but the separation of the energies of the states corresponding to individual values of M varies with the field strength. In the absence of a field, each energy level within a term, represented by the individual values of J, becomes degenerate with $(2J + 1)$ states.

At high field strengths, the coupling of L and S breaks down, and total orbital and spin angular momenta are individually space quantized to the field direction. The quantization is represented by the quantum numbers M_L and M_S, which take values from $-L, -L + 1, -L + 2, \ldots, L$ and $-S, -S + 1, -S + 2, \ldots, S$ respectively. Observationally, this decoupling of L and S results in the Paschen–Back effect (chapter 6).

3.4 TERM FORMATION

Since it can be a source of confusion, it is worth repeating certain definitions at this point. Energies within atoms may be described via Terms, Levels or States.

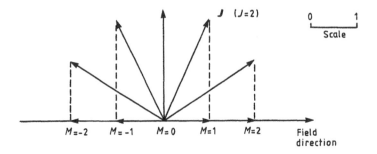

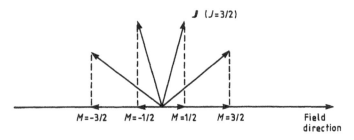

Figure 3.8   Space quantization of the total angular momentum of an atom in the presence of a magnetic field for $J = 2$ and $J = 3/2$.

An energy State is the energy corresponding to a specific value of the magnetic quantum number, M, for a particular value of J. Several states in general go to make a Level. In the absence of a magnetic field, such states all have the same energy. An energy Level is the energy corresponding to a specific value of J for a particular pair of values of L and S. Generally several levels with differing energies make up a term. Thus an energy Term is formed of the energy levels corresponding to a particular pair of values of L and S.

If the electrons contributing to L and S are non-equivalent (i.e. electrons belonging to different (n, l) subgroups), then the full range of terms obtained by all possible combinations of L and S is allowed. For equivalent electrons (same n and l), however, the Pauli exclusion principle excludes some possible combinations. Under this principle, as we have seen, if n and l have the same values for two electrons, then at least one of m_l and m_s for the electrons must differ.

Two non-equivalent electrons with $l_1 = l_2 = 1$ would have $L = 0, 1$ or 2 and $S = 0$ or 1, giving rise to the terms

$$L = 0 \quad S = 0 \quad J = 0$$
$$L = 0 \quad S = 1 \quad J = 1$$
$$L = 1 \quad S = 0 \quad J = 1$$
$$L = 1 \quad S = 1 \quad J = 2, 1, 0$$
$$L = 2 \quad S = 0 \quad J = 2$$
$$L = 2 \quad S = 1 \quad J = 3, 2, 1.$$

Two similar equivalent electrons, however, would be limited to the terms

$$L = 0 \quad S = 0 \quad J = 0$$
$$L = 1 \quad S = 1 \quad J = 2, 1, 0$$
$$L = 2 \quad S = 0 \quad J = 2.$$

The term $L = 2$ $S = 1$ $J = 3, 2, 1$ is not allowed because in order to have a value of L of 2, l_1 and l_2 must lie in same direction. The two electrons will therefore have identical values of m_l. Only the case with the spins antiparallel ($m_{s_1} = -1/2$ and $m_{s_2} = +1/2$) is therefore allowed by the exclusion principle. Thus $S = 0$ is allowed, but $S = 1$, which would require $m_{s_1} = m_{s_2} = -1/2$ or $m_{s_1} = m_{s_2} = +1/2$, is not. The reasons underlying the other restrictions are more complex and are given in full in Appendix B.

3.5 CLOSED SHELLS AND SUBSHELLS

In determining the whole atom quantum numbers from those of its constituent electrons, we have ignored contributions from closed shells and subshells. We are now in a position to justify this assumption. In a closed shell, all the electrons allowed by the Pauli exclusion principle for a given value of n are present. Similarly in a closed subshell all the allowed electrons for a given pair of values of n and l are present. Such electrons are equivalent, and must exist in antiparallel pairs with respect to their spins, or the shell would not be full. The value of S must therefore be zero. L is similarly zero as we may see by the trivial example of the lowest energy level, for which $l = 0$ for both electrons. Thus in closed shells and subshells we have $L = S = 0$, and their effect upon the whole atom quantum numbers may be ignored.

3.6 NOTATION

The various electron and whole atom quantum numbers are complex and difficult to follow. Unfortunately for the beginner, the situation is then rendered a hundred times worse by the illogical, inadequate and confusing system of notation that has been developed for the energy levels. That notation has developed historically from the observations of the early spectroscopists, and has its earliest origins from the time before there was any understanding of the structure of the atom. Since, however, all articles, books, tables, etc, that the student will encounter use this system of notation, he or she must perforce come to grips with it.

The notation is concerned with specifying the quantum numbers of a particular energy level in order to identify it uniquely. It is also concerned with providing information on which transitions (chapter 4) are allowed or forbidden. There are two related components to the notation, one for the individual electrons and one for the whole atom quantum numbers.

Let us deal with the individual electrons first. Since the notation system developed historically, it is helpful to look at its origins. The study of spectroscopy began by looking at spectra. Some elements, notably the alkalis, produced sets of lines of differing appearance or pattern. These sets, as noted previously, became labelled as Sharp, Principal, Diffuse and Fundamental. Subsequently, it was found that these corresponded to values of l of 0, 1, 2 and 3. Electrons with those values of l consequently have become called s, p, d or f electrons respectively. Electrons with values of l of 4 and upwards are similarly labelled by lower case letters, starting alphabetically from g. The principal quantum number, n, is just given by its numerical value. Thus the electrons in the ground state of iron, for example, are:

$$2 \times \text{1s electrons}$$
$$2 \times \text{2s electrons}$$
$$6 \times \text{2p electrons}$$
$$2 \times \text{3s electrons}$$
$$6 \times \text{3p electrons}$$
$$6 \times \text{3d electrons}$$
$$2 \times \text{4s electrons}$$

or, as it is more usually written, with the number of electrons as a following superscript:

$$1s^2, 2s^2, 2p^6, 3s^2, 3p^6, 3d^6, 4s^2.$$

For the whole atom quantum numbers, the value of L is indicated in a similar fashion to that of l, but using upper case letters. Thus S represents a value of L of 0, P represents a value of L of 1 etc. The value of S is indicated using the multiplicity, which, as we have seen, has the value $(2S + 1)$. This number is placed as a preceding superscript before the symbol for L.

Continuing with the energy notation, the value of J is shown as a following subscript. Finally a new term is introduced because of its significance in determining allowable transitions (chapter 4): the Parity. Parity is found from the sum of the individual values of l for the electrons. It is even or odd, accordingly as that sum is even or odd. In quantum mechanics the parity is found from whether or not the wavefunction of the term changes sign upon inversion. The parity is indicated in the symbol by a following superscript, 'o' when it is odd, and by the lack of such a superscript when it is even. Thus the whole atom quantum numbers for the ground state of iron are indicated by the symbol

$$^5D_4$$

The superscript '5' indicates a value of S of 2. The symbol 'D' indicates a value of L of 2. The subscript '4' indicates a value of J of 4. Finally, the lack of a following superscript shows the parity to be even. We may see this by looking at the values of l from the individual electrons: (above) we have values of l of $2 \times 0, 2 \times 0, 6 \times 1, 2 \times 0, 6 \times 1, 6 \times 2$ and 2×0, giving $\Sigma l = 24$, which is even.

The full symbol for the ground state of iron combines the symbols for the electrons with that for the atom:

$$1s^2, 2s^2, 2p^6, 3s^2, 3p^6, 3d^6, 4s^2, {}^5D_4.$$

However, as we have seen, the inner closed shells and subshells do not contribute to the whole atom quantum numbers, and so frequently only the details of the outer electrons are shown:

$$4s^2, {}^5D_4.$$

In a similar fashion we have the ground state of boron:

$$1s^2, 2s^2, 2p, {}^2P^o_{1/2}$$

or

$$2p, {}^2P^o_{1/2}$$

Here, the following superscript 'o' shows that the parity is odd ($2 \times 0 + 2 \times 0 + 1 \times 1 = 1$). Ionizing boron removes the outer electron, and so the ground state of ionized boron becomes

$$1s^2, 2s^2, {}^1S_0$$

and this is the same as the ground state of neutral beryllium, as would be expected.

The quantum numbers alter for energy levels higher than the ground state, and this appears in the symbols. Thus the ground state of helium is (now taking this as the zero level):

$$1s^2, {}^1S_0 \qquad (0\ eV)$$

and its first few excited energies are

$$1s, 2s, \ {}^3S_1 \qquad (19.82\,eV)$$

$$1s, 2s, \ {}^1S_0 \qquad (20.62\,eV)$$

$$1s, 2p, \ {}^3P^o_2 \qquad (20.96\,eV)$$

$$1s, 2p, \ {}^3P^o_1 \qquad (20.96\,eV)$$

$$1s, 2p, \ {}^2P^o_0 \qquad (20.96\,eV)$$

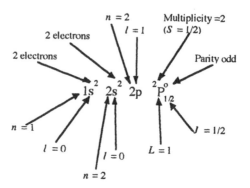

Figure 3.9   Ground state of boron.

Table 3.1   Terms for $l_1 = 3$ and $l_2 = 1$.

	$L = 4$	$L = 3$	$L = 2$
$S = 1$	${}^3G_{5,4,3}$	${}^3F_{4,3,2}$	${}^3D_{3,2,1}$
$S = 0$	${}^1G_4$	${}^1F_3$	${}^1D_2$

where the energies from the ground state are shown in electron volts after each symbol.

In our earlier notation, ${}^3P_2^o$ represents an individual energy level, made up of $5 \, (= 2J + 1)$ energy states. The three ${}^3P^o$ levels, with J values of 2, 1 and 0, make up the term.

We may summarize the information in the symbol as shown in figure 3.9.

For S terms, there is only one level because $L + S = |L - S|$ (equation (3.8)), but the multiplicity still continues to be used in the symbol.

For atoms and ions with two or more valence electrons, equations (3.6) to (3.8) are used. Thus for a p electron and an f electron, we have

$$l_1 = 3 \quad \text{and} \quad l_2 = 1 \qquad (3.9)$$

giving, from equation (3.6),

$$L = 4, 3 \quad \text{and} \quad 2 \qquad (3.10)$$

from equation (3.7),

$$S = 1, 0 \qquad (3.11)$$

and since $l_1 + l_2 = 4$, which is an even number, the parity is even. Thus we have the terms shown in table 3.1, where the individual J values for the levels making up the terms are also shown as subscripts.

Two non-equivalent p electrons similarly will have

$$l_1 = 1 \quad \text{and} \quad l_2 = 1 \qquad (3.12)$$

Table 3.2 Terms for $l_1 = l_2 = 1$ (non-equivalent electrons).

	$L = 2$	$L = 1$	$L = 0$
$S = 1$	$^3D_{3,2,1}$	$^3P_{2,1,0}$	$^3S_1$
$S = 0$	$^1D_2$	$^1P_1$	$^1S_0$

Table 3.3 Terms for $l_1 = l_2 = 1$ (equivalent electrons).

	$L = 2$	$L = 1$	$L = 0$
$S = 1$		$^3P_{2,1,0}$	
$S = 0$	$^1D_2$		$^1S_0$

giving

$$L = 2, 1 \text{ and } 0 \qquad (3.13)$$

as in the previous example,

$$S = 1, 0 \qquad (3.14)$$

and the parity will be even. The terms will thus be as in table 3.2.

With the non-equivalent electrons used in the examples above, the exclusion principle does not restrict the number of terms, since the electrons already have differing azimuthal quantum numbers.

For two equivalent p electrons, however, n and l will already be the same, and thus m_l and/or m_s must be different. If the two electrons have their l vectors in the same direction, giving a D term, then m_l is the same for both. The values of m_s must therefore differ. The triplet term, for which m_s for both electrons is the same, is therefore not allowed. Only the singlet term, for which one electron has $m_s = +1/2$ and the other $m_s = -1/2$ exists. Similarly, if the value of m_s for each electron is the same, giving a triplet term, then the m_l values must differ. The allowed terms for two equivalent p electrons are thus as in table 3.3.

For other pairs of equivalent and non-equivalent electrons similar tables can be drawn up. For more than two electrons, the number of terms increases sharply. Thus, for example, with two non-equivalent p electrons and one d electron, we have the terms given in table 3.4, where the figures in brackets indicate that the term specification can be obtained by more than one route. Such terms will in general have differing energies.

Extensive tables of allowed terms for various combinations of electrons may be found in the following:

Astrophysical Quantities by C W Allen (third edition, Athlone Press, 1973). Listings of the energies of individual levels within terms for specific atoms and ions may be found in

Table 3.4 Terms for $l_1 = l_2 = 1$ and $l_3 = 2$ (non-equivalent electrons).

	$L = 4$	$L = 3$	$L = 2$	$L = 1$	$L = 0$
$S = 1.5$	^4G	$^4F(2)$	$^4D(3)$	$^4P(2)$	^4S
$S = 0.5$	$^2G(2)$	$^2F(4)$	$^2D(6)$	$^2P(4)$	$^2S(2)$

Atomic Energy Levels volumes 1, 2 and 3 by C E Moore (US National Bureau of Standards, NSRDS-NBS 35, 1971)
A Multiplet Table of Astrophysical Interest by C E Moore (NBS technical note 36, 1959)
An Ultra-violet Multiplet Table by C E Moore (NBS Circular 488, 1950)
Lines of the Chemical Elements in Astronomical Spectra by P W Merrill (Carnegie Institute of Washington publication 610, 1958).
and in the references given in
Bibliography on Atomic Transition Probabilities (1914 through October 1977) by J Fuhr, B Miller and G Martin (NBS Special Publication 505, 1978) and Supplement 1 (NBS Special Publication SP 505-1, 1980)
Bibliography on Atomic Energy Levels and Spectra (NBS Special Publication 363, 1980) and Supplements 1, 2 and 3 (NBS Special Publications SP 363 S1 1971, SP 363 S2 1980, SP 363 S3 1985).

3.7 HYPERFINE STRUCTURE

Many spectral lines are found to be made of a number of closely spaced components when observed at very high resolution. Typically separations of the components are less than 0.01 nm. These components imply that individual energy levels are split in a manner not explained by the foregoing analysis. This splitting of the levels arises from interactions of the nucleus and the electrons. Protons and neutrons, like the electrons, belong to the group of subatomic particles that have a half-integer spin and are known as Fermions. The nuclei of atoms therefore possess spins resulting from those of their component particles. Since the nucleus is also electrically charged, it has a associated magnetic moment. The value of the nuclear magnetic moment is about 1/2000 that for an electron because of the much greater mass of nuclei.

Previously the total angular momentum of the atom has been denoted by J. For most astrophysical purposes this is an adequate approximation because the nuclear magnetic moment is so much smaller than that from the electrons. Nonetheless, we may now see that J is strictly just the total angular momentum of the electrons in the atom, and that the true total angular momentum of the atom must also take account of the angular momentum of the nucleus. The angular momentum of the nucleus is denoted by I, and the true total angular momentum of the atom by F. The associated quantum number, F, following previous practice, takes the values

$$F = (J + I), (J + I - 1), (J + I - 2), \ldots, |J - I|. \qquad (3.15)$$

The J and I vectors precess around the F vector (cf space quantization above) and produce different energies for the different values of F. The magnetic moment of the nucleus, however, is so much smaller than that of the electrons that the energy differences for different values of F are very small. Thus for hydrogen, for which $I = 1/2$, in the ground state ($J = 1/2$) we have $F = 1$ or 0, and the two levels differ in energy by only 5.9×10^{-6} eV.

The separation of spectral lines produced by hyperfine splitting of the energy levels is so small that for astronomical sources it is usually swamped by other line broadening effects. Hyperfine splitting is thus not of much general importance to the astrophysicist with the exception of the molecular radio lines. There is one notable exception to this rule for the atoms, however, and that is the hydrogen 21 cm line. This line is widely used for the study of galaxies through its emission by hydrogen atoms in interstellar gas clouds (molecular hydrogen is very difficult to observe, see chapter 5). The 21 cm line results from a transition between the two hyperfine levels in the ground state of hydrogen ($F = 1$ to $F = 0$).

3.7.1 Isotope effect

A second effect, which is sometimes included within the term hyperfine structure because spectral line separations comparable to those due to the nuclear magnetic moment are found, is due to the different masses of the nuclei of different isotopes of the same element. The mass of the nucleus affects the reduced mass (equation (2.8)), and thus the energy levels (equation (2.7)) of the atom. For the lighter elements, the separation of lines due to different isotopes may be sufficiently large to be observed in astronomical sources. The relative abundances of different isotopes can then be studied, and this can have considerable importance. Thus the abundance of heavy hydrogen (deuterium) has been determined observationally and is a significant constraint on models of the early stages of the 'Big Bang'.

4

Transitions

4.1 BASICS

We have seen how electrons in atoms interact to determine the energy levels. If an atom changes from one such level to another, then in general its total energy will alter. Such a change is usually called a Transition. Transitions are often thought of as electrons moving within the atom from one orbit to another. However, as we have seen, in many cases it is the relative interactions of the orbits which change between levels, with the individual electrons not changing their properties. Howsoever it occurs, if the energy of an atom changes, then either it must gain the required energy from some source when the energy increases, or lose it when the energy reduces. Energy may be gained from a variety of sources: from thermal energy during collisions between atoms, from radioactive decay, from chemical reactions, from electrical and magnetic fields etc. However, in most circumstances the energy is gained from the radiation field by the absorption of a photon. It is this last process, of course, which is of interest to spectroscopists. In theory, if energy is to be lost from an atom during a transition, then it could be absorbed by the inverse of any of the above processes for supplying energy. Most of the time, however, except in thermodynamic equilibrium, the energy will be lost by the emission of a photon.

The removal of a photon from the radiation field by a transition in which the atom gains energy (usually called an upward transition) alters that radiation field. If the field is then studied by a spectroscope (chapter 8), then the change to it could theoretically be detected. In practice the loss of one photon would not be noticed amongst the normal fluctuations in the arrival of photons at the detector. However, if many atoms are undergoing the same transition, then they will all subtract photons of the same energy (and therefore wavelength; equation (1.4)), from the radiation field, because the difference in energy between the two levels will be the same for all the atoms. Then the absence of these many photons will show up in a spectroscope as a dark line at the appropriate wavelength. (Remember the appearance of the dark portion of the spectrum as a *line* only occurs because of the linear entrance slit to most spectrographs—see chapter 1.)

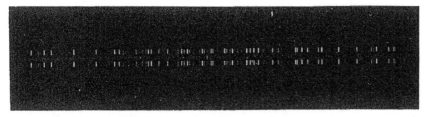

Figure 4.1   An emission line spectrum.

Figure 4.2   An absorption line spectrum.

Such dark lines are usually called Absorption Lines because they arise from the absorption of photons by atoms during upward transitions.

A transition from a higher to a lower energy in an atom is usually called a downward transition. It results in the addition of a photon to the radiation field. Again, if many atoms are undergoing the same downward transition, then sufficient photons may be added to the radiation field to show up in the spectrum as an Emission Line.

A photon may be emitted irrespective of the nature of the surrounding radiation field. There may even be no other photons, in which case the emission is an isolated feature in the spectrum against a dark background. Such a spectrum, for example, would be found for a sodium street light or an illuminated neon advertising sign, though in both these cases there would be many more than one emission line, since many different transitions are occurring in the sodium and neon atoms (figure 4.1).

For an upward transition to occur, however, the radiation field must clearly contain a least one photon of the required energy. In practice, an absorption line is usually found silhouetted against a bright continuous spectrum. Such a continuous spectrum is normally produced by a hot solid, liquid or dense gas (see also the discussion in chapter 1). In a continuous spectrum, there is emission of photons over a wide range of wavelengths, with the intensity of the emission varying only slowly with wavelength. The appearance of an absorption spectrum is thus normally that of a bright background with one or more dark lines across it (figure 4.2).

Transitions can occur between different energy levels or, in the presence of magnetic or electric fields, between the individual states comprising different levels. Looking at the energy levels of actual atoms (e.g. figure 3.2), it is apparent that the number of different transitions if we just take all possible combinations of pairs of levels is going to be very large indeed. Yet many atoms have relatively simple spectra with small numbers of lines. Thus the Balmer spectrum of hydrogen can be calculated to a high degree of accuracy by the simple Rydberg formula (equation (1.3)). The reason for this simplicity is that not all the possible transitions are allowed to occur. The rules determining which transitions can occur are complex. They are called Selection Rules and are discussed in the next section.

4.2   SELECTION RULES

If we look at the Grotrian diagram for hydrogen and hydrogen-like ions (figure 3.1), then it would appear that the first line in the Balmer series (Hα; wavelength 656.279 nm) can result from 15 different transitions, i.e. from any of the $n = 2$ levels (figure 4.3):

$$2s\,^2S_{1/2}, \; 2p\,^2P^0_{1/2}, \; 2p\,^2P^0_{3/2}$$

to any of the $n = 3$ levels (figure 4.3):

$$3s\,^2S_{1/2}, \; 3p\,^2P^0_{1/2}, \; 3p\,^2P^0_{3/2}, \; 3d\,^2D_{3/2}, \; 3d\,^2D_{5/2}.$$

In practice, it is found that only seven of the transitions do occur (figure 4.4). This restriction arises from the operation of three selection rules. The first is that L must change by 0 or ± 1, the second that J can only change by 0 or ± 1 and the third that the parity must change. Thus any transitions from S to S, P to P, or S to D terms are forbidden, which eliminates seven of the possibilities. The final forbidden transition is from

$$2s\,^2P^0_{1/2} \quad \text{to} \quad 3d\,^2D_{5/2}$$

which would require J to change by $+2$, i.e. from 1/2 to 5/2.

We can thus divide transitions into Allowed and Forbidden. Some forbidden transitions are found to occur under special conditions (see later discussion), but their probability of occurrence in the laboratory or in stellar spectra is normally many orders of magnitude smaller than that of allowed transitions. For many, but not all, astronomical purposes we can therefore consider only the allowed transitions.

The selection rules were found empirically at first. Later they were shown to be derivable from the conservation of angular momentum when the quantum numbers are large. At large quantum numbers, quantum theory must coincide with classical theory. If we then assume that the rules so derived for large quantum numbers also apply to small quantum numbers (sometimes called

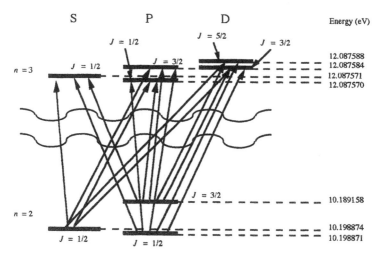

Figure 4.3 Partial Grotrian diagram for hydrogen showing all the Hα transitions possible in the absence of selection rules.

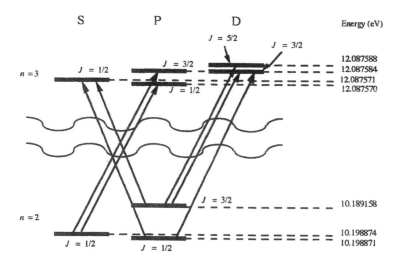

Figure 4.4 Partial Grotrian diagram for hydrogen showing Hα transitions allowed by the selection rules.

Bohr's Correspondence Principle), then the selection rules follow, for all transitions.

For our purposes, it is sufficient to know what the selection rules are, in order to be able to apply them, without worrying about their derivation. For normal transitions under L–S coupling, the selection rules are thus

$$\Delta J = 0, \pm 1 \quad (J = 0 \ddagger J = 0)$$
$$\Delta S = 0$$
$$\Delta L = 0, \pm 1 \quad (L = 0 \ddagger L = 0)$$
Parity changes

where the symbol $\ddagger$ indicates that a particular combination is not allowed. In addition, in the presence of a magnetic field (etc), we have the rule $\Delta M = 0, \pm 1$, and for hyperfine splitting, $\Delta F = 0, \pm 1$ with $F = 0 \ddagger F = 0$.

When a single electron jumps, Δn is unrestricted and Δl is ± 1. With more than one valence electron it is possible, though rare, for two electrons to jump simultaneously with the absorption or emission of a single frequency. In such a case, the value of Δn for either electron is again unrestricted, but $\Delta l_1 = \pm 1$, and $\Delta l_2 = 0, \pm 2$.

Thus the following transitions are allowed:

$$1s\,^2S_{1/2} \rightarrow 2p\,^2P^0_{3/2}$$
$$3p^2\,^4P_{5/2} \rightarrow 3d\,^4P^0_{3/2}$$
$$4p\,^3S_1 \rightarrow 5s\,^3P^0_1$$
$$3p^4\,^3P_0 \rightarrow 4s\,^3S^0_1$$

while these are not:

$$2s\,^2S_{1/2} \ddagger 3s\,^2S_{1/2} \quad \text{(parity unchanged)}$$
$$2s\,^2S_{1/2} \ddagger 3d\,^2D_{3/2} \quad (\Delta l = 2,\ \Delta L = 2,\ \text{parity unchanged})$$
$$2s^2\,^1S_0 \ddagger 2p\,^3P^0_2 \quad (\Delta J = 2,\ \Delta S = 1)$$
$$4s\,^4P_{1/2} \ddagger 2p^3\,^2D^0_{3/2} \quad (\Delta S = 1).$$

The set of spectral lines resulting from all the allowed transitions between two terms is called a Multiplet. The lines within a multiplet will not all be of the same intensity, and a guide to their relative intensities can be obtained from theoretical considerations. Thus for example, the

$$^3P \rightarrow\ ^3D$$

multiplet has six transitions between individual levels, with predicted relative strengths as shown in table 4.1. More complete tables for such relative strengths

Table 4.1 Relative strengths of lines within a multiplet.

	$^3D_3$	$^3D_2$	$^3D_1$
$^3P_2$	21	3.75	0.25
$^3P_1$		11.25	3.75
$^3P_0$			5

Table 4.2 Relative strengths between multiplets.

Multiplet	Relative intensity
$2p^2\,^3P \rightarrow 2p\,2d\,^3D^0$	135
$2p^2\,^3P \rightarrow 2p\,2d\,^3P^0$	45

may be found in *Astrophysical Quantities,* page 61ff (C W Allen, 3rd edn 1973, Athlone Press) and in the references therein.

Absorption lines originating from the ground state of an atom or ion are usually very strong. Thus, for example, the H and K Fraunhofer lines in the solar spectrum originate from the ground state of ionized calcium. Such lines are known as resonance lines, and their strength is due to the very large majority of atoms or ions being in their ground state even at temperatures of several thousand degrees (see the Boltzmann formula, equation (4.5)).

Relative intensities between multiplets can also be predicted (the intensity of a multiplet is given by the sum of the individual intensities of its lines). This is relatively straightforward for multiplets arising from the same electron jump (related multiplets). Thus, the multiplets arising from one of two valence electrons jumping from a p to a d orbit have the relative intensities shown in table 4.2.

More complete tables of relative intensities between multiplets are to be found in *Astrophysical Quantities*, p 66ff (reference above). The relative strengths of unrelated multiplets are much more difficult to calculate; again details are given in *Astrophysical Quantities*.

Knowledge of such predicted relative intensities is very useful when trying to identify unknown lines in a spectrum. If one such line is tentatively identified, then the other lines in the same multiplet should be present in approximately their theoretical relative intensities. Thus if the initial identification is of a weak line in a multiplet and the stronger lines are not found, then the identification is probably wrong. Conversely, if several lines from the multiplet can be found, then the identification stands a good chance of being correct. The calculated intensities should not be relied on completely, however, since the populations of the excited levels in atoms can vary a good deal from those of thermodynamic equilibrium. Thus in low density environments such as interstellar nebulae, stellar coronae etc, metastable levels (levels from which there are only forbidden downward transitions) can become overpopulated compared with the populations in thermodynamic equilibrium, leading to forbidden lines (see later in this chapter) dominating the spectrum. In the laboratory, excitation of lines by electric sparks or arcs can lead to unusual population balances between different levels. Nonetheless, if used cautiously, the predicted relative strengths within multiplets are a useful tool to the spectroscopist.

In our discussion of selection rules, we have so far assumed that L–S coupling (chapter 3) can be applied. This assumption is valid most of the time in astronomical spectroscopy. Nonetheless, there are occasions when $j-j$ coupling becomes important. Since L and S no longer have any meaning under this form of interaction, the selection rules applying to them disappear. The selection rules under $j-j$ coupling thus become

$$\Delta j_i = 0, \pm 1$$
$$\Delta J = 0, \pm 1 \quad (J = 0 \updownarrow J = 0).$$

Now pure $j-j$ coupling occurs relatively seldom, and within a single atom some terms may remain determined by $L-S$ coupling while others are influenced by $j-j$ coupling. Thus, in practice, the importance of $j-j$ coupling is in modifying the $L-S$ coupling selection rules, rather than having a widespread application in its own right. The commonest such modification results in the relaxation of the $L-S$ selection rule, $\Delta S = 0$, so that transitions in which the multiplicity changes can occur. Thus for example in the spectrum of neutral silicon, we may find lines arising from the transition

$$3p^2\,^3P \rightarrow 3p\,4s\,^1P^0$$

where S changes from 1 to 0. Lines arising from such transitions are forbidden by the strict $L-S$ coupling selection rules, and so are often called Forbidden Lines. This particular breakdown of $L-S$ coupling due to the onset of $j-j$ coupling is a relatively mild and quite prevalent transgression against the $L-S$ selection rules. Lines resulting from it are often regarded as not 'really' forbidden, and are called Intercombination Lines. Intercombination lines can often be found in normal stellar spectra, though their intensities will usually be much less than those of comparable allowed lines and multiplets. The presence of intercombination lines in the spectrum of an atom indicates that strict $L-S$ coupling is starting to break down. When writing down transitions and their resulting lines, a system of square-bracket notation is used to indicate the nature of the transitions (the positioning of the brackets can vary):

No brackets: allowed line, e.g. H I 656.3 or O I 630.0
One bracket: intercombination line, e.g. Mg I] 457.1 or K III 348.1]
Two brackets: forbidden line, e.g. [N III] 386.9 or [O II 732.0].

What we may now call 'true' forbidden lines arise from other modes in which $L-S$ coupling fails. This may occur in several ways. The selection rule itself may only be a first approximation, the transition may occur through magnetic dipole or electric quadrupole interactions, or the presence of electric or magnetic fields from nearby atoms and ions may disturb the $L-S$ coupling sufficiently for the transition to take place. These variations allow changes such as $\Delta J = \pm 2$, $\Delta L = \pm 2$, $\Delta M = \pm 2$ to occur during transitions, or for the parity not to change. In the laboratory or in stellar spectra, lines arising from such forbidden transitions may be expected to be from 10^{-5} to 10^{-12} times fainter than the allowed lines in the same atom.

4.3 TRANSITION PROBABILITIES

The rather loose descriptions of relative intensities in lines and multiplets used in the previous section need to be more precisely defined if they are to be used in quantitative spectroscopy. That more precise definition is via Transition Probabilities at the atomic level, and via Absorption and Emission Coefficients at the macroscopic level. The actual strengths of lines in a star's spectrum will also depend upon the element's abundance, stellar temperature, rotational velocity, magnetic fields etc, and these latter factors are considered in subsequent chapters.

There are three transition probabilities associated with every transition, and these are commonly called the Einstein transition probabilities. Conceptually, we may most easily approach transition probabilities through a second quantity, the Lifetime of an excited energy level. We have seen (above) that some transitions are allowed and others forbidden, and that amongst the latter, some transitions are 'more forbidden' than others. This variation in degree of allowability of a transition may be quantified through the average time it would take an atom excited to the upper level of the transition spontaneously to drop to the lower level. That time is called the lifetime of the transition. The lifetime of the level is the average time an electron would remain in it when all downward transitions are included, and is the inverse of the sum of the transition probabilities. Lifetimes range from around 10^{-8} s for allowed transitions, through perhaps 10^{-5} s for intercombination transitions, to upwards of 10^{-3} s for forbidden transitions. The forbidden transition producing the well-known 21 cm hydrogen line, for example, has a lifetime of 11 million years!

The first of the transition probabilities, called the Spontaneous Emission Transition Probability, is just the reciprocal of the lifetime of the transition. From the concept of the lifetime, we may therefore see that the spontaneous transition probability is the probability that the atom will spontaneously undergo the downward transition in unit time. This transition probability is customarily symbolized as A_{21}, where 2 and 1 are the upper and lower levels of the transition, respectively. From the values for the lifetimes, we may see that A_{21} takes values from $10^8 \, \text{s}^{-1}$ for allowed transitions to $10^{-15} \, \text{s}^{-1}$ or less for the most extreme of the forbidden transitions. The number of spontaneous transitions per unit time from a unit volume is thus given by

$$N_2 \, A_{21} \qquad (4.1)$$

where N_2 is the number density of atoms excited to the upper level.

The second transition probability is the Absorption Transition Probability. This is clearly different in nature from the spontaneous transition probability because, whatever its value, no transitions will occur in the absence of radiation. The actual probability of a transition occurring therefore depends upon the intensity of the radiation at the appropriate wavelength, as well as the properties of the atom. The number of absorptions per unit volume and time is thus

$$N_1 \, B_{12} \, I_{21} \tag{4.2}$$

where B_{12} is the Absorption Transition Probability from the lower level to the upper level, N_1 is the number density of atoms in the lower level and I_{21} is the intensity of radiation with the wavelength appropriate to induce the transition from level 1 to level 2 (i.e. that emitted by a transition from level 2 to level 1).

The final transition probability is sometimes called that of Negative Absorption. More commonly, however, the process is called Stimulated Emission. Nowadays, the existence of stimulated emission is familiar through the ubiquitous laser (which name, remember, is an acronym for Light Amplification by Stimulated Emission of Radiation). Such familiarity, however, disguises a process that many students find to be a difficult concept. It may therefore be useful to look at the process from a classical point of view for a moment.

Consider the pendulum (an oscillator) of a mechanical clock (figure 4.5). In normal circumstances, it receives an impulse from the escapement once per oscillation. That impulse is in phase with the oscillation and adds energy to the pendulum. In the absence of frictional energy losses etc, the pendulum would increase its amplitude of oscillation. In other words, the oscillator (pendulum) absorbs energy from the 'wave' (escapement impulse). However, if we were to alter the escapement so that its impulse were to oppose the motion of the pendulum (i.e. 180° out of phase), then the pendulum's motion would die down, and its energy would be added to the escapement train of the clock. In this last case we thus have the oscillator adding energy to the 'wave'. In more general terms, when an oscillator and a wave interact, the exchange of energy is from the wave to the oscillator when their phase difference is between ±90°, and from the oscillator to the wave when the phase difference is outside that range. The first situation corresponds to normal absorption, the second to stimulated emission.

Stimulated emission can thus only occur in the presence of radiation of the same wavelength as that produced by the transition. Regarding it as negative absorption, we may see from equation (4.2) that the number of stimulated emissions per unit volume and time is

$$N_2 \, B_{21} \, I_{21} \tag{4.3}$$

where B_{21} is the Stimulated Emission Transition Probability from the upper level to the lower level.

The photons produced by stimulated emission are added to the radiation field with the same direction, polarization and phase as those of the stimulating photons. Photons emitted spontaneously, by contrast, have random directions, polarizations and phases. Examples of some transition probabilities are shown in table 4.3.

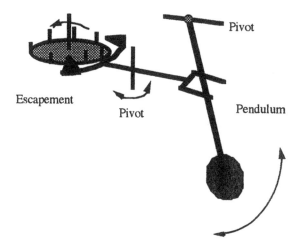

Figure 4.5   Schematic view of the escapement and pendulum of a mechanical clock.

Table 4.3   Examples of transition probabilities (units of s^{-1}).

Line	A_{21}	B_{21}	B_{12}
H I 121.5 (Ly-α)	6.2×10^8	2.8×10^{12}	1.4×10^{12}
H I 102.5 (Ly-β)	1.7×10^8	4.5×10^{11}	2.2×10^{11}
H I 97.2 (Ly-γ)	6.8×10^7	1.6×10^{11}	7.9×10^{10}
O I 394.7	3.7×10^5	5.7×10^{10}	8.0×10^{10}
Mg I 457.1]	2.1×10^2	5.1×10^7	1.7×10^7
[N I 519.8]	1.6×10^{-5}	5.7	5.7

4.3.1   Inter-relationships

The three transition probabilities are related to each other. In order to follow that relationship, we need a new concept, that of Statistical Weight.

When we have a large number of identical atoms in thermodynamic equilibrium (when a single temperature characterizes all physical processes), then the proportion excited to a given energy (the population of that energy) depends only upon the energy and the temperature. Now we saw in chapter 3 that an energy level is in general composed of a number of energy states. Except in the presence of a magnetic or electric field, these states have identical energies. Each state is associated with a value of the magnetic quantum number, M, which takes the values $-J, -J + 1, -J + 2, \ldots, J - 1, J$. Thus there are $(2J + 1)$ states in a level. Since each such state has the same energy as all the others, the populations of each state are also identical. The population of the level, made up from adding together the populations of its states, is thus proportional to the number of states comprising it. The number of states in a level is called the Statistical Weight, g, of that level, and is given by

$$g = 2J + 1 \qquad (4.4)$$

(the statistical weight of an individual state is 1). If two different levels have exactly the same energies, then their relative populations will be in the ratio of their statistical weights.

The population of an excited level varies exponentially with temperature, and inverse exponentially with energy. Thus the relative population of two levels within a species of atom, in thermodynamic equilibrium, is given by

$$\frac{N_2}{N_1} = \frac{g_2}{g_1} e^{-(E_2-E_1)/kT} \qquad (4.5)$$

where E_2 is the energy of the upper level etc, and k is Boltzmann's constant ($= 1.38062 \times 10^{-23}$ J K^{-1}). This last equation is generally known as Boltzmann's formula and is of widespread use.

The concept of statistical weight is also extended to energy terms, though this is not strictly valid because the levels within the term may have different energies. If the energy differences between the levels are small, however, then the statistical weight of the term is given by $(2S+1)(2L+1)$, and is the number of energy states in all the levels within the term.

To return to our discussion of the relationships between the transition probabilities, under conditions of thermodynamic equilibrium, we have the Principle of Detailed Balancing. This states that every process is balanced by its inverse. In this case it requires every upward transition between two levels to have a downward transition occurring nearby and nearly simultaneously. Thus in thermodynamic equilibrium we must have the total number of absorptions equal to the total number of emissions:

$$N_1 \, B_{12} \, I_{21} = N_2 \, B_{21} \, I_{21} + N_2 \, A_{21}. \qquad (4.6)$$

In thermodynamic equilibrium the radiation field is that of a black body and the intensity is given by the well known Planck equation,

$$I(v, T) = \frac{2hv^3 \mu^2}{c^2 (e^{hv/kT} - 1)} \qquad (4.7)$$

or

$$I(\lambda, T) = \frac{2hc^2 \mu^2}{\lambda^5 (e^{hv/\lambda kT} - 1)} \qquad (4.8)$$

where μ is the refractive index of the medium (normally close to unity), $I(v, T)$ and $I(\lambda, T)$ are the intensity of the radiation at frequency v or wavelength λ, and per unit frequency or wavelength interval, at temperature T. (NB: it is common practice to use the symbols $B(v, T)$ and $B(\lambda, T)$ for these quantities, the symbol B coming from Black body. That practice is avoided here because of the ease of confusion with the symbols for the transition probabilities, B_{12} and B_{21}.)

Now by substituting into equation (4.6) for I_{21} from equation (4.7) and for N_2/N_1 from equation (4.5), we may find

$$A_{21} = \frac{2g_1 h\nu^3 \mu^2}{g_2 c^2} B_{12} \frac{e^{h\nu/kT} - g_2 B_{21}/g_1 B_{12}}{e^{h\nu/kT} - 1}. \qquad (4.9)$$

Since the transition probabilities are properties of the atom, and are independent of temperature, the second ratio in equation (4.9) must also be independent of temperature. This can only occur if

$$\frac{g_2 B_{21}}{g_1 B_{12}} = 1 \qquad (4.10)$$

and so we get

$$g_2 B_{21} = g_1 B_{12} \qquad (4.11)$$

From equation (4.10) we thus see that the second ratio in equation (4.9) has a value of unity, and so we finally obtain

$$A_{21} = \frac{2g_1 h\nu^3 \mu^2}{g_2 c^2} B_{12} = \frac{2h\nu^3 \mu^2}{c^2} B_{21}. \qquad (4.12)$$

From equation (4.12) we may see that for the two emission transition probabilities,

$$\frac{A_{21}}{B_{21}} \propto \nu^3 \qquad (4.13)$$

and so at high frequencies (infrared and upwards) stimulated emission is usually negligible.   Observationally this is confirmed by the detection of naturally occurring masers in interstellar gas clouds, but not of the higher frequency lasers.

4.4   ABSORPTION AND EMISSION COEFFICIENTS

The Einstein transition probabilities determine the probabilities of an individual transition occurring, and thus of there being a resulting absorption or emission line in the spectrum.   So far we have regarded energy levels as though their energies were precisely defined.   Were this actually to be the case then such absorption or emission lines would appear at precise wavelengths and be of zero width in wavelength terms.   In fact, energies of levels are not precisely determined, but as a result of the Heisenberg uncertainty principle they spread over a small range of energy, centred upon the nominal value.

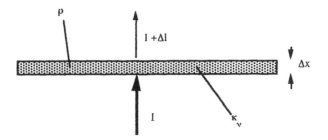

Figure 4.6 The reduction in the intensity of a beam of radiation on passing through a thin absorbing layer of density ρ and thickness Δx.

The Heisenberg uncertainty principle is normally stated as

$$\Delta x \Delta p_x \approx h/2\pi \qquad (4.14)$$

where x is a positional coordinate, p_x the momentum in the same direction, and h is Planck's constant. Now

$$p_x = m v_x \qquad (4.15)$$

and so

$$\Delta x \Delta p_x = 2(\Delta x/v_x)(0.5\Delta m v_x^2) = 2\Delta t \Delta E$$

where ΔE is the uncertainty in the energy of the level and Δt is (roughly) the lifetime of the level under emission transitions. Hence the principle, as it applies to atomic energy levels, takes the form

$$\Delta t \Delta E \approx h/4\pi. \qquad (4.16)$$

Therefore in a transition between two levels, both of which now have a certain 'fuzziness', the resulting energy change must also have a certain range of values and the ensuing spectral line will no longer be sharp.

Since we now see that the spectral line must cover a certain range of wavelengths, the Einstein coefficients, which have no wavelength dependence, are clearly inadequate to describe the situation completely. We therefore introduce the concepts of Absorption and Emission Coefficients, which do have a wavelength dependence.

The Absorption Coefficient per unit mass, κ_v, is defined in terms of the effect of a thin layer of an absorbing material upon a beam of radiation passing through it (figure 4.6). The change in intensity, ΔI, is clearly related to the original intensity, I, by

$$\Delta I = -I\kappa_v\rho\Delta x. \qquad (4.17)$$

When the layer is thick, we may find its effect by summing the contributions from the thin layers comprising it, or in the limit as Δx tends to zero,

$$\int_{I_1}^{I_2} \frac{\mathrm{d}I}{I} = \int_{x_1}^{x_2} -\kappa_v\rho \, \mathrm{d}x \qquad (4.18)$$

where I_1 is the intensity at level x_1 etc. Hence, provided that the absorption coefficient and density are independent of x,

$$I_2 = I_1 e^{-\kappa_\nu \rho (x_2 - x_1)}.\qquad(4.19)$$

The emission coefficient, ε_ν, is more simply related to the emitted radiation, in the absence of absorption, by

$$I_2 = I_1 + \varepsilon_\nu \rho (x_2 - x_1).\qquad(4.20)$$

4.4.1 Natural broadening

Processes that lead to spectral lines having a finite width are usually called Broadening Mechanisms. There are numerous such mechanisms, and those of astrophysical significance, such as Doppler Broadening, Pressure Broadening, the Zeeman Effect etc, will be encountered later (chapter 13) for the information that they reveal about the material in which the lines originate. The broadening that results from the intrinsic widths of energy levels, however, is a property of the atom, and is usually called Natural Line Broadening. As we shall see shortly, an important quantity in widespread use, called the Oscillator Strength, compares the actual atom with its classical equivalent; it is therefore useful to start by looking at the absorption coefficients for a classical oscillator. In the case of weak lines the variation of these coefficients also gives the shape of the spectral line, or as it is more usually known, the line profile.

The mass absorption coefficient for a classical oscillator is given by

$$\kappa_\nu = \frac{Ne^2}{4\pi\varepsilon_0 \rho mc} \frac{\gamma/4\pi}{(\nu - \nu_0)^2 + (\gamma/4\pi)^2}\qquad(4.21)$$

where N is the number of absorbing oscillators per unit volume, m and e are the mass and charge of the oscillator (assumed to be an electron) ν_0 is the resonant frequency (i.e. the central frequency of the absorption line) and γ is the damping constant for the oscillation caused by radiation from the oscillator, and given by

$$\gamma = \frac{2\pi e^2 \nu^2}{3\varepsilon_0 mc^3}.\qquad(4.22)$$

The variation of absorption coefficient with frequency is shown in figure 4.7.

We may expand equation (4.19) to find the relationship between the line profile and the absorption coefficient:

$$I_2 = I_1 \left(1 + \frac{-\kappa_\nu \rho \Delta x}{1!} + \frac{(-\kappa_\nu \rho \Delta x)^2}{2!} + \ldots \right)\qquad(4.23)$$

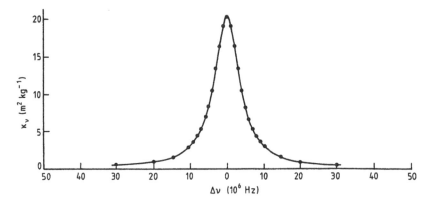

Figure 4.7 Variation of absorption coefficient with frequency due to classical natural broadening for Hα ($\lambda = 656$ nm, $\nu = 4.57 \times 10^{14}$ Hz).

where Δx has been written for $(x_2 - x_1)$. For small values of $(-\kappa_\nu \rho \Delta x)$ we have therefore

$$I_2 \approx I_1(1 - \kappa_\nu \rho \Delta x) \tag{4.24}$$

and the line profile for weak absorption lines has the same shape as the variation of κ_ν with frequency, as already mentioned. In the absence of absorption, an emission line will also have the same profile. The shape of this line profile is often called the Lorentz or Damping profile.

A commonly used parameter in observational spectroscopy is the half-width of the line. This is the width of the line at half its central depth (figure 4.8). Its use is often preferred to that of the actual (or whole) width of the line because the latter can be difficult to determine in the presence of noise. From equation (4.21) we may see that the second ratio on the right hand side has a value of $(1/(\gamma/4\pi))$ when $\nu = \nu_0$, and has a value of $0.5(1/(\gamma/4\pi))$ when $(\nu - \nu_0) = \gamma/4\pi$. Thus the absorption coefficient, and hence from equation (4.24), the line intensity, has half its peak value when

$$\nu - \nu_0 = \gamma/4\pi \tag{4.25}$$

i.e.

$$\Delta\nu_{1/2} = 2(\nu - \nu_0) = \gamma/2\pi \tag{4.26}$$

where $\Delta\nu_{1/2}$ is the half-width of the line in frequency terms.

From equation (4.22) therefore,

$$\Delta\nu_{1/2} = \frac{e^2 \nu^2}{3\varepsilon_0 m c^3}. \tag{4.27}$$

Since

$$\Delta\lambda = -c\nu^{-2}\Delta\nu \tag{4.28}$$

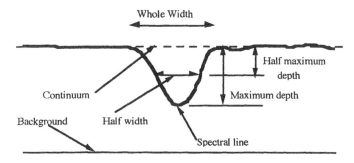

Figure 4.8 The definition of whole and half-widths for an absorption line.

we also have the half-width in wavelength terms,

$$\Delta\lambda_{1/2} = \frac{e^2}{3\varepsilon_0 mc^2} \tag{4.29}$$

and we see the remarkable result that this is a constant whose value is 1.18×10^{-14} m ($0.000\,011\,8$ nm). Thus classically, for weak lines, all naturally broadened lines have the same line profile in wavelength terms.

Observationally, the natural linewidth is rarely directly significant because it is far smaller than the linewidths produced by other broadening mechanisms, and so its effects are swamped. For absorption in the interstellar medium, however, where the density is very low and there are no other significant broadening mechanisms, the resonance line of hydrogen, Lyman-α at 122 nm, has its shape determined almost entirely by natural broadening.

In the quantum mechanical version of natural broadening, a new quantity appears which is of major importance. This is the Oscillator Strength, f, mentioned previously. In quantum mechanics, a similar expression to equation (4.21) for the absorption coefficient is obtained but modified in two ways. Firstly the damping constant (relabelled Γ) is related to the lifetimes of the levels and is given by

$$\Gamma = \Gamma_1 + \Gamma_2 = \frac{1}{T_1} + \frac{1}{T_2} \tag{4.30}$$

where Γ_1 is the damping constant for level 1 etc and T_1 is the lifetime for level 1 etc. The value of Γ for permitted transitions is thus around 10^8, comparable to that of γ for visual transitions. We have previously seen that the lifetime for a level under spontaneous transitions is just the reciprocal of the spontaneous transition probability. Now, however, we must consider the lifetimes under all possible transitions, not just the spontaneous ones, and so we get

$$\Gamma_1 = \frac{1}{T_1} = \sum A_{1n} + \sum B_{1n} I_{1n} + \sum B_{1m} I_{1m} \tag{4.31}$$

Table 4.4 Oscillator strengths for hydrogen.

Hα	0.637
Hβ	0.119
Hγ	0.044
Hδ	0.021
Hε	0.012
⋮	⋮
Total for all the absorption lines	0.866
Contribution from ionizations	0.238
Emission line (Lyman-α)	−0.104
Sum (equals number of electrons in hydrogen)	1.000

where n represents the levels lower than level 1 (remember level 1 in the sense used here and earlier is the lower level of the transition, not necessarily the ground state), m represents the levels higher than level 1, I_{1n} is the intensity for radiation corresponding to the transition level 1 to level n etc and the summation in each case is over all available levels. An analogous equation gives Γ_2.

The second modification is that N, the number of absorbing oscillators per unit volume, is replaced by Nf. Here, f is a correction factor known as the Oscillator Strength which converts the classical values to the quantum mechanical values. The oscillator strength may be envisaged as the number (usually fractional) of classical oscillators that would produce the observed line strengths. Oscillator strengths may be defined for absorption or emission, and if the emission oscillator strengths are regarded as negative, then the sum of all the oscillator strengths over all possible transitions from a level within an atom or ion will equal the number of electrons in that atom or ion (a relationship usually known as the Thomas–Reiche–Kuhn Sum Rule).

For hydrogen-like ions and a few other simple atoms, oscillator strengths may be calculated. Usually, however, they have to be measured experimentally, and uncertainties in their values remain a major problem for determinations of element abundance (chapter 14). Tables of f-values and the related transition probabilities (see below) may be found in the references and bibliographies given near the end of chapter 3.

For the Balmer series of hydrogen, for example, the oscillator strengths are as shown in table 4.4.

With these two modifications, the expression for the natural absorption coefficient becomes

$$\kappa_\nu = \frac{Nfe^2}{4\pi\varepsilon_0\rho mc} \frac{\Gamma/4\pi}{(\nu - \nu_0)^2 + (\Gamma/4\pi)^2}. \tag{4.32}$$

The half-width of the line is easily seen, by the previous argument, to be $\Gamma/2\pi$, but it does now depend upon the lines' identity through equation (4.31).

4.4.2   More inter-relationships

The absorption and emission oscillator strengths for a pair of levels are simply related to each other by the statistical weights of the levels:

$$g_1 f_{12} = -g_2 f_{21}. \tag{4.33}$$

The oscillator strengths are also related to the transition probabilities (see above). Thus

$$A_{21} = \frac{2\pi v_{21}^2 e^2}{\varepsilon_0 mc^3} f_{21} \tag{4.34}$$

and so from equations (4.12) and (4.33)

$$B_{21} = \frac{\pi e^2}{\varepsilon_0 v_{21} hmc} f_{21} \tag{4.35}$$

$$B_{12} = \frac{\pi e^2}{\varepsilon_0 v_{21} hmc} f_{12}. \tag{4.36}$$

We thus see that the oscillator strength is a more fundamental quantity upon which the Einstein transition probabilities all depend. The relationship between oscillator strength and line strength that this leads to will be employed in later chapters.

4:4.3   Relationship between transition probabilities and absorption and emission coefficients

We may rewrite equation (4.32) as

$$\kappa_v = \frac{Ne^2}{4\pi \varepsilon_0 \rho mc} f\phi(v) \tag{4.37}$$

where $\phi(v)$ represents the second term on the right hand side of equation (4.32), and is called the shape function for the natural line profile. Other shape functions may be substituted when other broadening mechanisms, such as pressure, predominate. Now from equation (4.36) this equation becomes

$$\kappa_v = \frac{v_{21} h}{4\pi^2 \rho} N_1 B_{12} \phi(v) \tag{4.38}$$

However, as we have seen, stimulated emission is negative absorption, and so κ_v must be reduced to take account of this:

$$\kappa_v = \frac{v_{21} h}{4\pi^2 \rho} \phi(v) (N_1 B_{12} - N_2 B_{21}). \tag{4.39}$$

The form of the absorption coefficient given by equation (4.39) is what is normally intended by the term 'Absorption Coefficient', and it will be used in that sense henceforth. The quantity given by equations (4.32), (4.38) etc is the absorption coefficient uncorrected for stimulated emission and will be so referred to in future.

The emission coefficient may be found similarly, by assuming that the absorption shape coefficient also fits the emission distribution (normally this is a good assumption). We thus get

$$\varepsilon_\nu = \frac{\nu_{21}h}{4\pi^2\rho} \phi(\nu)N_2A_{21}. \tag{4.40}$$

The ratio of absorption and emission coefficients may be seen to be

$$\frac{\varepsilon_\nu}{\kappa_\nu} = \frac{N_2A_{21}}{N_1B_{12} - N_2B_{21}} \tag{4.41}$$

and so dependent only upon the temperature (because N_1 and N_2 vary with T). This is a fundamental result known as Kirchhoff's law and one which we have already encountered (equation (1.1)).

4.5 IONIZATION AND RECOMBINATION

The transitions with which we have been concerned so far have been between specific energy levels within the atom or ion. Such levels have a maximum energy which is called the ionization limit (see figure 3.1 for example). If an electron acquires more energy than this limit, then it becomes detached from the atom or ion as a free electron and leaves behind an ion or a more highly ionized ion. Such transitions are often called Bound–Free because the electron goes from being *bound* to the atom to being *free* of it. Transitions wherein the electron remains attached to the atom or ion are called similarly, Bound–Bound, while those in which the electron is free both before and after the transition are Free-Free. The latter are an important class of transitions in astrophysics, and are discussed in more detail in chapter 6.

A commonly used notation to indicate the degree of ionization of an atom is to append a Roman numeral after the chemical symbol. The value of the Roman numeral is one more than the number of electrons that the neutral atom has lost. Thus neutral iron is Fe I, singly ionized iron is Fe II, doubly ionized iron is Fe III, etc. Thirteen-times ionized iron would be Fe XIV. The notation Fe^+, Fe^{++}, etc is also used, where the number of pluses gives the number of electrons that the atom has lost, but this is cumbersome for high levels of ionization. The latter form of notation however is widely used to indicate the negative hydrogen ion, H^-. The positive charge of the hydrogen nucleus is not completely screened by its normal single electron. The neutral atom can

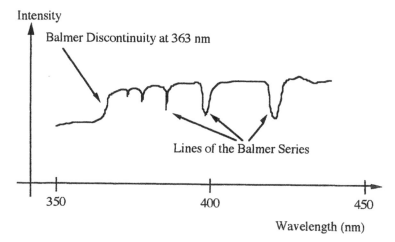

Figure 4.9 The Balmer Discontinuity: the absorption edge due to ionization of hydrogen from the $n = 2$ level.

therefore attract and weakly hold a second electron. The second electron has only two energy levels in the ion, and these have ionization limits of 0.76 and 0.29 eV respectively. The ion is of great importance for stellar spectra, since for solar-type stars the opacity in the optical region is mostly due to the negative hydrogen ion.

Ionization usually requires a significant amount of energy. Thus hydrogen requires 13.6 eV for ionization, and helium 24.6 eV for its first ionization and another 54.4 eV to be doubly ionized. To remove the last electron from 14-times ionized phosphorus requires 3070 eV! This compares with 1 to 5 eV for typical bound–bound transitions. The spectral features resulting from ionization and its opposite effect, Recombination, are thus to be found from the radio to the ultraviolet and even x-ray parts of the spectrum (see below). Since the electron's energy above the ionization limit is effectively unquantized, ionization produces an absorption band rather than a line (figure 4.9). Such bands are called ionization edges, and are of major importance for the ultraviolet spectra of stars, especially for ionization from the ground state of the particle (Resonant Ionization).

Ionization can occur through the absorption of a single photon of sufficient energy, or through collisions etc. It is also possible for ionization to occur without direct absorption of energy. This can happen when other processes have excited two electrons simultaneously within the atom or ion. One of those electrons can then drop down to its ground state. Instead of a photon being emitted during that transition, however, the energy is taken by the second excited electron, and if sufficient, raises it above the ionization limit. This effect is known as Autoionization or Preionization, or if the energies involved are in the x-ray region, as the Auger effect. The inverse effect, wherein an

Table 4.5   The Balmer decrement.

Hα	2.79
Hβ	1.00
Hγ	0.473
Hδ	0.262
etc	

electron recombines with an ion but excites another electron rather than emitting a photon, is known as Dielectronic Recombination.

When an electron recombines with an ion, it may do so to any of the energy levels in that ion. Unless the recombination is directly to the ground state, the electron will cascade down through the excited levels, emitting various spectral lines, on its way to the ground state. This process is a special case of Fluorescence in which longer wavelength lines are emitted as the result of a shorter wave absorption (figure 1.6). The emission lines from recombinations are of considerable significance for astrophysics, especially for interstellar nebulae. In such nebulae the photons are emitted only through spontaneous transitions since the radiation field bathing the atom or ion is exceedingly weak. Collisional and other processes are also generally negligible and so the relative intensities of the lines are determined solely by the properties of the atom. In the case of hydrogen, the changing intensities of the Balmer lines are known as the Balmer Decrement. For interstellar nebulae, it should thus take the values relative to Hβ (table 4.5).

4.6   X-RAY SPECTRA

Lines arising from electronic transitions usually lie in the optical part of the spectrum. As the heavier atoms become progressively more and more ionized, however, the remaining electrons are more and more tightly bound to the nucleus. Transitions among such remaining electrons therefore generally involve much larger energy changes than those we have considered up to now. The resulting lines will be found at very short wavelengths as well as in the optical region. Thus, for example, 13-times ionized iron (Fe XIV), which occurs in the solar corona, can only be further ionized by radiation of wavelength 3.16 nm or shorter. Many of its spectral lines lie at only slightly longer wavelengths. Conventionally, the region from about 0.1 to 100 nm is called the x-ray region, and transitions from highly ionized atoms therefore produce x-ray spectral lines. There is little difference in principle between transitions of these inner electrons, producing x-ray lines, and those with which we are more familiar, producing optical lines. The detectors available to astronomers for observations in the x-ray and gamma ray regions, however, do not have sufficient energy resolution at present to study x-ray lines from astronomical objects, and so the topic is not pursued further here.

5

Spectra of Molecules

5.1 INTRODUCTION

Molecular spectroscopy is a huge and complex subject and to do it justice would require many books far longer than this one. As astronomers (rather than physical chemists), however, our interest in the topic is quite specialized. Thus in stars we need only concern ourselves with simple diatomic molecules like TiO and CH. More complex molecules may be found in interstellar space, H_2CO or HCCCN etc, but then we are primarily concerned only with their radio transitions. The spectroscopy of solids is of importance for interstellar dust particles, and for planetary surfaces, but as we shall see, it can be of only very limited help in those studies. Planetary atmospheres probably involve the most opportunity for the molecular spectroscopist, but even there the main molecular species encountered, H_2, CO_2, N_2 etc, are moderately simple. Nonetheless, even with this restricted brief, only an outline of the topic can be covered here, and the reader is referred to the bibliography for more detailed works. Many of the molecular transitions produce lines at long wavelengths. They are therefore strictly outside the brief for this book. Nonetheless such transitions are covered for completeness and because their omission would lead to a distorted view of molecular processes.

Molecular spectroscopy is more complex than atomic spectroscopy for one main reason: molecules themselves are more complex and offer many more degrees of freedom, and therefore energy states, than do atoms. The energy levels due to the configuration of electrons in molecules are much more numerous than in atoms, and so the spectral lines resulting from transitions between them are also far more numerous; so much so that often the lines overlap and blend (except in very high dispersion spectra) leading to the appearance of Bands rather than lines in spectra (figure 5.1).

The increased complexity, however, does not end with the larger number of electron energy levels: a molecule can also possess energy in two forms not available to an atom—rotation and vibration. Energy in these latter two forms is quantized and transitions between energy levels can occur as for the electron

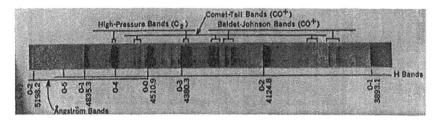

Figure 5.1 Optical band spectra due to molecules. Reproduced from *Molecular Spectra and Molecular Structure. Vol 1 Spectra of Diatomic Molecules* by G Herzberg (New York: Van Nostrand Reinhold).

energy levels and lead to spectral lines †.

The lines resulting from these new forms of energy, however, appear in quite different parts of the spectrum from those involving electrons (figure 5.2). Transitions between electron energy levels typically involve energy changes of 1 to 10 eV, and the lines appear in the near infrared, visual and near ultraviolet parts of the spectrum (a region often usefully called the Optical part of the spectrum). Transitions due to changes in vibrational energy levels involve energy changes of only about 1% of those for the electrons, and the resulting lines appear in the infrared. Changes a hundred times smaller still are typical for rotational transitions, and the lines are then in the microwave part of the spectrum.

These energy differences arise through the differences in masses between electrons and nuclei and scale roughly as

$$\frac{E_e}{E_v} \approx \sqrt{\frac{m_n}{m_e}} \quad \text{and} \quad \frac{E_e}{E_r} \approx \frac{m_n}{m_e} \tag{5.1}$$

where E_e is the electronic energy, E_v is the vibrational energy, E_r is the rotational energy, m_n is the mass of the nucleus and m_e is the mass of the electron.

The disparity in velocities between nuclei and electrons due to their mass differences enables a considerable simplification of the understanding of molecules to be made. This is known as the Born–Oppenheimer Approximation. By this approximation we may regard the wavefunction for the molecule as being composed of three separate independent functions, one each for the electrons,

---

† A fourth quantized form of energy exists for particles due to their motions through space (sometimes called their translational energy). The energy levels involved, however, are so close to each other that they effectively form a continuum. Transitions between such levels are only of spectroscopic significance for electrons and produce Free–Free or Bremsstrahlung radiation (chapter 6). The levels do become of significance under conditions of extreme density such as may be found in white dwarfs and neutron stars, when together with the operation of the Pauli exclusion principle they result in electron and neutron degeneracy pressure.

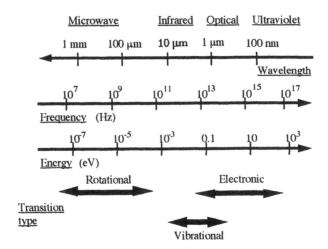

Figure 5.2 Approximate ranges for molecular transitions.

vibrations and rotations, and dependent upon electron configuration, internuclear separation, and rotation angle respectively. The energy of the molecule is then just the simple sum of the energies of each mode:

$$E = E_e + E_v + E_r. \tag{5.2}$$

We may thus consider the energy levels, transitions and spectral lines for molecules arising from electrons, vibrations and rotations almost separately from each other. The word 'almost' occurs in the last sentence because a vibrational transition has so much more energy than a rotational transition that a process inducing a vibrational transition will almost certainly change the molecule's rotational characteristics as well. Thus we may consider rotational transitions in isolation, but we have to consider vibrational and rotational transitions together. In a similar way electronic transitions will normally also involve vibrational and rotational changes.

5.2 ROTATIONAL TRANSITIONS

The rotational energy levels can have transitions between themselves, resulting in lines in the microwave and radio parts of the spectrum, but they also produce fine structure within lines arising from the vibrational levels. This latter effect is considered in the next section.

Vibrations within molecules occur so much faster than rotations, as we may see from the respective frequencies of their lines (figure 5.2), that we may consider rotation as that of a molecule whose internuclear distances are fixed at the time-averaged values over the period of a vibration. Thus we may imagine

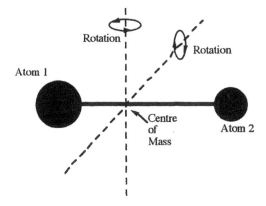

Figure 5.3 Diatomic molecule as a rigid rotator.

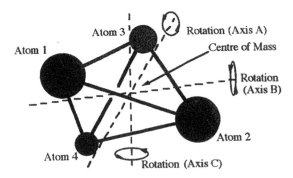

Figure 5.4 A polyatomic (in this example, 4-atom) molecule as a rigid rotator.

a diatomic molecule as a rigidly rotating dumb-bell (figure 5.3). It will clearly have two significant rotational degrees of freedom, since the moment of inertia around the axis joining the nuclei is zero and so rotations around it will have zero energy. With three or more atoms, unless they are all exactly in a straight line, like CO_2, all three axes of rotation can be involved in rotational transitions (figure 5.4). The concept of the molecule as a rigid rotator is normally a good one. It can, however, break down at very high rotational speeds when centrifugal distortion may stretch the bonds slightly. The effect of this is just to reduce the energy of the rotational level by a small amount.

The solution to the Schrödinger equation for a rigid rotator is straightforward in the simple case shown in figure 5.3 and leads to the angular momentum, p, being quantized as

$$p = \frac{h}{2\pi}\sqrt{J(J+1)} \tag{5.3}$$

where J is the molecular total angular momentum quantum number (cf the total angular momentum or inner quantum number for an atom which conventionally

Table 5.1 Rotational
energy levels for the
lowest state of TiO.

J	E (eV)
0	0
1	1.26×10^{-4}
2	3.78×10^{-4}
3	7.56×10^{-4}
4	1.26×10^{-3}

uses the same symbol). The energy corresponding to a value of J is simply found from the classical formula for rotational energy (remembering that $p = I\omega$, where I is the moment of inertia and ω the angular velocity)

$$E = \frac{I\omega^2}{2} = \frac{h^2 J(J+1)}{8\pi^2 I}. \tag{5.4}$$

As in the atomic case, the energy level corresponding to a specific value of J is degenerate, and in the presence of a magnetic or electric field splits into $(2J + 1)$ states.

For the diatomic molecule shown in figure 5.3, the moment of inertia is

$$I = \frac{m_1 m_2}{m_1 + m_2} d^2 \tag{5.5}$$

where m_1 and m_2 are the masses of nuclei 1 and 2, and d is their mean internuclear separation. Thus for the lowest electronic and vibrational energy state of titanium oxide (TiO), one of the first molecules to appear in stellar spectra, which has a mean internuclear distance of 0.162 nm and nuclear masses of 47.9 and 16.0 amu, we find, as an example, the rotational energy levels shown in table 5.1 and figure 5.5.

The diatomic molecule is the simplest case, and has values for the three moments of inertia, I_A, I_B, and I_C, about the three perpendicular axes (figure 5.4) of

$$I = \frac{m_1 m_2}{m_1 + m_2} d^2 = I_A = I_B \quad \text{and} \quad I_C = 0 \tag{5.6}$$

(it is not critical which axis is labelled A, B or C). Other linear molecules have rotational energy level values given by equation 5.4 when the appropriate moment of inertia is used for I.

In a couple of other cases, symmetry allows the rotational energy levels to be directly calculated. Thus 'symmetric tops' (figure 5.6), wherein $I_A = I_B \neq I_C$ and $I_C \neq 0$, have their energy given by

$$E = \frac{h^2 J(J+1)}{8\pi^2 I_A} + \left(\frac{h^2}{8\pi^2 I_C} - \frac{h^2}{8\pi^2 I_A} \right) K^2 \tag{5.7}$$

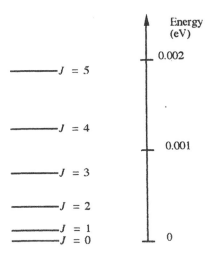

Figure 5.5 Rotational energy levels for the lowest electronic and vibrational energy level of TiO.

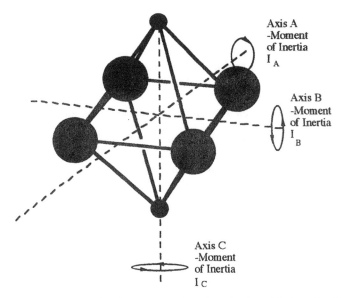

Figure 5.6 An example of a 'symmetric top' molecule with $I_A = I_B \neq I_C$.

where K is a second quantum number. 'Spherical tops', such as CH_4, with all three moments of inertia equal to each other and non-zero, are covered by equation (5.4).

In more general cases the molecule will have all three moments of inertia different and will have very complex rotational energy levels that cannot be calculated from simple formulae.

5.2.1 Selection rules

Particles and radiation interact only through the effects of the radiation's oscillating electric and magnetic fields on the particle. Generally it is the electric component that is important (though some forbidden line transitions arise through magnetic interactions, chapter 4). Thus either the electron configuration of the particle must change during the interaction, or the particle must be an electric dipole. The dipole arises when the centre of charge for the electrons is spatially separated from that of the nuclei. But in the microwave region wherein the rotational transitions lie, the energy of the radiation is insufficient to change the electron configuration, so interactions can only be via the dipole. Thus we obtain the first selection rule for molecular rotational transitions: the molecule must have an electric dipole moment.

Now homonuclear diatomic molecules like H_2, O_2 and N_2 have no electric dipole moment. Similarly more complex but symmetrical molecules like the linear CO_2 and spherical tops like CH_4 (see above) also have zero dipole moments. These and other molecules without dipole moments therefore do not have allowed rotational spectra (or, as we shall see later, vibrational spectra either)—a matter of some inconvenience to the astrophysicist studying the interstellar medium where the bulk of the material is in the form of H_2! (though forbidden lines from H_2 arising through quadrupole effects are detected).

The second selection rule is analogous to that for J within atomic transitions, and is

$$\Delta J = +1 \text{ on absorption} \qquad \Delta J = -1 \text{ on emission} \qquad (5.8)$$

which we may understand as arising through the photon involved in the interaction carrying one unit of angular momentum. For the more complex molecules, the additional quantum number, K, has the selection rule that it does not change during a rotational transition.

5.2.2 Transitions

During rotational transitions, the molecule is changing from one rotational state to another. Now some rotational levels have identical energies (i.e. they are degenerate), but most of the time the change will alter the energy of the molecule. Just as for atoms therefore, if the energy decreases during the change, then a photon will be emitted, while absorption of an appropriate photon causes an increase in energy. In each case, the frequency or wavelength of the photon is related to the energy change (given by the difference in energy levels) by equation (1.4).

Thus from equations (5.4) and (5.8), we have for heteronuclear diatomic molecules

Table 5.2 Frequencies and wavelengths of the first few spectral lines due to pure rotational transitions in the lowest electronic and vibrational level of TiO.

J (lower level)	J (upper level)	Frequency (Hz)	Wavelength (mm)
0	1	3.05×10^{10}	9.84
1	2	6.10×10^{10}	4.92
2	3	9.14×10^{10}	3.28
3	4	1.22×10^{11}	2.46
4	5	1.52×10^{11}	1.97

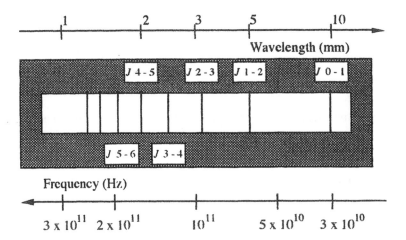

Figure 5.7 Sketch of the first few spectral lines due to pure rotational transitions in the lowest electronic and vibrational level of TiO.

$$\Delta E = \frac{h^2(J + 1)}{4\pi^2 I} \qquad (5.9)$$

where J is the quantum number for the lower level involved in the transition.

The resulting spectrum is a series of lines equally spaced in frequency (figure 5.7 for example). The centrifugal bond stretching at high values of J causes the lines to converge slowly but the effect does not normally become significant for J values of less than a 100.

For symmetrical tops, a formula similar to equation (5.9) may be found using equation (5.7). While no simple formula exists for the complex molecules, the same principle that the photon energy equals the difference between two energy levels still applies.

For the previously given example of TiO, the first few pure rotational spectral lines are listed in table 5.2, and sketched in figure 5.7.

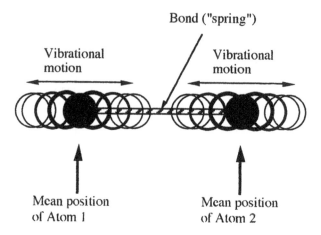

Figure 5.8   Vibrational motion in a diatomic molecule.

5.3   VIBRATIONAL TRANSITIONS

As already mentioned, a transition involving a change in the vibration of a molecule requires so much more energy than that involved in changing the molecule's rotation that the latter usually also occurs. We do not normally therefore have pure vibrational transitions, but ones which involve changes in both vibration and rotation. Thus the reader will often find such changes elsewhere referred to as Vibrational–Rotational transitions. The rotational changes generally have just the effect of introducing fine structure into the spectral lines arising from vibrational transitions, so here we retain the term Vibrational Transition.

A good approximation to the molecule as a vibrator is to imagine the bonds replaced by springs obeying Hooke's law, and the atoms by point masses. In a diatomic molecule we may therefore only have vibrations along the axis joining the nuclei (figure 5.8). More generally, if a molecule contains N atoms, then it will have $3N$ degrees of freedom. Three of these degrees of freedom correspond to motion in the three Cartesian coordinates, and another three to rotation about those axes. Thus a molecule will normally have $(3N - 6)$ modes of vibration. A linear molecule can only rotate about two axes, and therefore will have $(3N - 5)$ modes of vibration. Thus the diatomic molecules referred to above have the $(3 \times 2 - 5) = 1$ modes of vibrations already seen. A linear triatomic molecule will have four modes of vibration (figure 5.9), and clearly the range of possible vibration modes will become very large as the number of atoms increases.

The vibrations may be classified according to their type, as follows (see also figure 5.10)

ν vibrations: change of length of one or more bonds while angles remain constant (minimum number of bonds = 1).

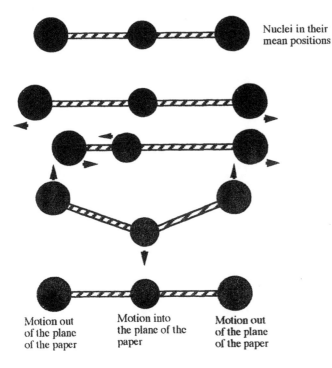

Nuclei in their
mean positions

Motion out
of the plane
of the paper

Motion into
the plane of the
paper

Motion out
of the plane
of the paper

Figure 5.9 The four modes of vibration for a linear triatomic molecule. The nuclei vibrate about their mean positions. The sketch shows instantaneous positions and motions.

δ vibrations: change of bond angle while bond lengths remain constant, i.e. a planar bend (minimum number of bonds = 2)

γ vibrations: out of plane bending, i.e. one atom oscillates through a plane defined by at least three other atoms. Both bond lengths and angles change (minimum number of bonds = 3).

τ vibrations: the angle between two planes each defined by at least three atoms, but with a bond in common, changes, i.e. a torsional or twisting change. Bond lengths and angles between adjacent atoms are constant but between more distantly connected atoms both change (minimum number of bonds = 3).

Normally the frequencies of these vibrations increase in the order; $\tau < \gamma < \delta < \nu$. Larger molecules can have even more complex modes of vibration, but these are not usually of importance for the astrophysicist.

5.3.1 Energy levels

From the previous outline, it may be anticipated that the vibrational energy levels are going to be complex in the extreme. This is certainly true in theory.

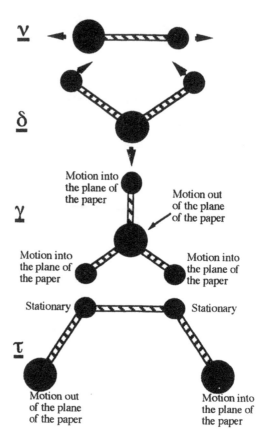

$\underline{\nu}$

$\underline{\delta}$

$\underline{\gamma}$

Motion into
the plane of
the paper

Motion out
of the plane
of the paper

Motion into
the plane of
the paper

Motion into
the plane of
the paper

Stationary

Stationary

$\underline{\tau}$

Motion out
of the plane
of the paper

Motion into
the plane of
the paper

Figure 5.10 Classification of modes of vibration of molecules. The nuclei vibrate about their mean positions. The sketch shows instantaneous positions and motions.

Fortunately the situation is rescued somewhat in practice for the astrophysicist by the simplicity of the molecules of importance. More generally the situation is simplified because normally only the transitions between the ground and first excited vibrational levels are of any significance (see selection rules below).

The quantization of vibrational energies for molecules may be investigated by solving the Schrödinger wave equation as before. A very simple but still useful approach for diatomic molecules, however, is by analogy with a vibrating string constrained at each end. Then, as is well known, we will have the fundamental vibration and its harmonics (figure 5.11), and these will correspond to the allowed vibrational energy levels.

This analogy has to be corrected in three ways. Firstly, the increasing energy (i.e. higher harmonics) corresponds to greater amplitudes for the molecular vibrations, and so the separation of the constraining points for the vibrating

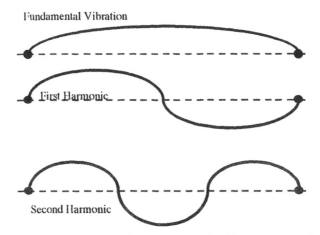

Figure 5.11 Vibrations of a string constrained between two points.

string has to be increased. A harmonic oscillator satisfies this requirement, and is a good model around the minimum energy point (figure 5.12). In a real molecule, however, the force required to push the nuclei together increases sharply at small internuclear distances, and so the actual curve will rise more steeply than that of a true harmonic oscillator. Thirdly, if the amplitude of the vibrations becomes large enough, the molecule will dissociate. In our model this corresponds to the outer constraint moving out to infinity at a finite value for the vibrational energy (figure 5.12). The envelope of the energy levels shown in figure 5.12 is approximated by the behaviour of an anharmonic oscillator, and is given by an equation of the type

$$E_s = (E_0 + E_D)\left(1 - e^{-a(s-s_e)}\right)^2 \tag{5.10}$$

where E_s is the potential energy for an internuclear separation of s, E_0 is the zero point energy (because of the Heisenberg uncertainty principle the minimum vibrational energy is not zero, but is $(h\nu_{vib}/2)$, where ν_{vib} is the fundamental frequency of the oscillation). E_D is the molecular dissociation energy from the vibrational ground state. s_e is the equilibrium internuclear separation. a is a constant determining the rate of approach of the potential to its asymptotic value.

The actual values of the vibrational energy levels shown in figure 5.12 are found from the solution of the Schrödinger equation for an anharmonic oscillator, and are given to a good degree of accuracy by

$$E_v = h\nu_{vib}\left[(v + \tfrac{1}{2}) - x(v + \tfrac{1}{2})^2\right] \tag{5.11}$$

where v is the vibrational quantum number and takes values $0, 1, 2, \ldots$, and x is a small correction to allow for the anharmonicity; it ensures that the separation

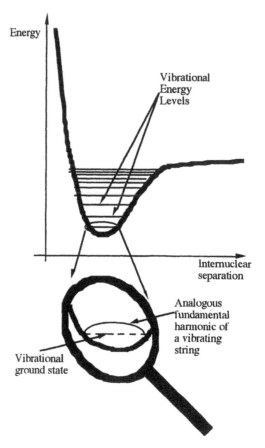

Figure 5.12   Vibrational energy levels.

between successive energy levels decreases towards the dissociation limit as shown in figure 5.12.

Thus for TiO, for which ν_{vib} has the value 3.025×10^{13} Hz, and x is 0.00427, we get the first few energy levels as listed in table 5.3.

Without the anharmonic term (x), the energy intervals would be constant at 0.120 eV. The effect of the anharmonicity is thus to halve the energy intervals by the time we reach $v = 50$.

5.3.2   Selection rules

We have the first requirement, as for rotational interactions, that the molecule must possess an electric dipole if it is to interact with the radiation. Thus homonuclear diatomic and other zero-dipole molecules do not have vibrational

Table 5.3 The energies of the first few vibrational levels of TiO.

v	Energy (eV)
0	0.060
1	0.179
2	0.297
3	0.413
4	0.529

spectra. However, weak spectra due to higher order and magnetic interactions may sometimes be found, so allowing interstellar molecular hydrogen to be observed via its $v = 2$ to $v = 1$ and $v = 1$ to $v = 0$ emission lines near 2 μm.

A second requirement is that the dipole moment should change during a vibrational transition.

The third selection rule is analogous to that for J within rotational transitions, and is

$$\Delta v = +1 \text{ on absorption} \qquad \Delta v = -1 \text{ on emission} \qquad (5.12)$$

which we may again understand as arising through the photon involved in the interaction carrying one unit of angular momentum. We may also have $\Delta v = 0$, but this of course just corresponds to the pure rotational transitions already considered.

Finally, for the lowest electronic level ($\Lambda = 0$ see below), we have that the rotational quantum number J must change. For higher electronic levels, however, it can remain constant, i.e.

$$\Delta J = \pm 1 \quad \text{for } \Lambda = 0 \qquad (5.13)$$
$$\Delta J = 0, \pm 1 \quad \text{for } \Lambda > 0 \qquad (5.14)$$

The change in J can be ± 1 for either emission or absorption now (cf equation (5.8)) because the majority of the energy change is taken up by the vibrational transition.

These selection rules strictly apply to the harmonic oscillator approximation. The effect of anharmonicity is to allow Δv to be greater than 1. Such changes, however, have a low probability, and any resulting lines will be very weak.

5.3.3 Transitions

Of the selection rules just discussed, the most significant is that in equation (5.12). For most molecules of interest in astronomy, their temperature is a few hundred Kelvin or less. Boltzmann's formula (equation (4.5)) then tells us that for a first excited level at around 0.1 eV (see table 5.3 for example), the population of the excited level will be 0.1% or less that of the ground state. Only at stellar temperatures will the higher vibrational levels therefore have significant populations. Thus, except in the latter case, we are concerned only with the transition $v = 0$ to $v = 1$ ($v = 0$ to $v = 2$ etc being forbidden by equation (5.12)).

The rotational levels within a single vibrational level mean that within the limits of the various selection rules, a vibrational transition can have many components. If we consider absorption from the lowest electronic level and the $v = 0$ vibrational level, then this must be to the $v = 1$ level. Within that transition, we can have rotational changes of $\Delta J = \pm 1$. The fundamental frequency arising from the transition from $v = 0$ $J = 0$ to $v = 1$ $J = 0$ therefore is not allowed, and we have a series of lines with higher and lower frequencies. By convention, the lines with $\Delta J = -1$ belong to what is termed the P branch, while those with $\Delta J = +1$ belong to the R branch (figure 5.13). The energies may be obtained from equations (5.2), (5.4) and (5.11):

$$\Delta E_{vr} = h\nu_{vib}(1 - 2x) + \frac{h^2}{8\pi^2 I_1} J_1(J_1 + 1) - \frac{h^2}{8\pi^2 I_0} J_0(J_0 + 1) \qquad (5.15)$$

where the subscript 1 refers to vibrational level $v = 1$ and subscript 0 refers to vibrational level $v = 0$.

For small values of J, to a good approximation $I_1 \approx I_0$, and so remembering that for the R branch we have $J_1 = J_0 + 1$, while for the P branch we have $J_1 = J_0 - 1$, we get the simplified formulae

$$\Delta E_{vr}(R) = h\nu_{vib}(1 - 2x) + \frac{h^2(J + 1)}{4\pi^2 I} \qquad (5.16)$$

for the R branch, while for the P branch

$$\Delta E_{vr}(P) = h\nu_{vib}(1 - 2x) - \frac{h^2 J}{4\pi^2 I} \qquad (5.17)$$

where in each case J is the value of the rotational quantum number for the $v = 0$ level. The spectral lines therefore form equally spaced groups on either side of ν_{vib} for small values of J (figures 5.14 and 5.15). The spacing becomes non-uniform as J increases, because the moment of inertia increases through the centrifugal stretching of the internuclear distance. This has the most dramatic effect on the R branch, causing the separations to decrease first to zero, and then to negative values. On a Fortrat diagram (figure 5.16), we may see that we get a clustering of R branch spectral lines around the maximum frequency. This clustering is called the Band Head. For most practical purposes, however, we may ignore the centrifugal effect, because the intensity of the spectral lines becomes negligible for J greater than about 10.

The energies and corresponding frequencies and wavelengths are listed for TiO in table 5.4.

Table 5.4 The rotational fine structure lines for the vibrational transition from $v = 0$ to $v = 1$ for the lowest electronic level of TiO.

J for the $v = 0$ level	J for the $v = 1$ level	Energy (eV)	Frequency (Hz)	Wavelength (μm)	
ν_{vib} no spectral line produced					
0	0	0.11889	2.9992×10^{13}	10.003	
R branch					
0	1	0.11902	3.0025×10^{13}	9.9917	R0
1	2	0.11914	3.0055×10^{13}	9.9817	R1
2	3	0.11927	3.0088×10^{13}	9.9708	R2
3	4	0.11940	3.0121×10^{13}	9.9598	R3
4	5	0.11952	3.0151×10^{13}	9.9499	R4
P branch					
1	0	0.11876	2.9959×10^{13}	10.014	P1
2	1	0.11864	2.9929×10^{13}	10.024	P2
3	2	0.11851	2.9896×10^{13}	10.035	P3
4	3	0.11838	2.9863×10^{13}	10.046	P4
5	4	0.11826	2.9833×10^{13}	10.056	P5

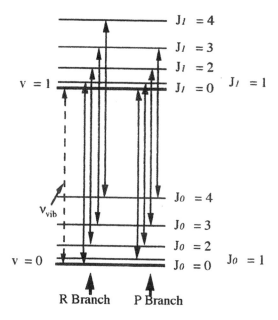

Figure 5.13 Rotational fine structure of the first vibrational transition for the lowest electronic level in a diatomic molecule.

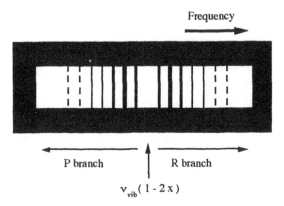

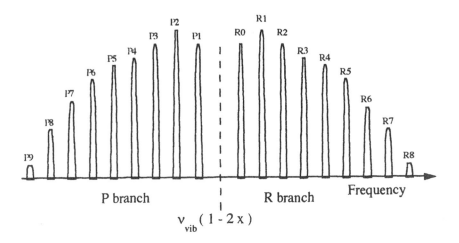

Figure 5.14 Sketch of the absorption spectrum for a vibrational transition from $v = 0$ to $v = 1$ for the lowest electronic level of a diatomic molecule.

Figure 5.15 Sketch of a tracing of an emission spectrum for a vibrational transition from $v = 0$ to $v = 1$ for the lowest electronic level of a diatomic molecule.

5.4 ELECTRONIC TRANSITIONS

5.4.1 Nomenclature

For diatomic molecules the energy levels arising from differing configurations of the electrons bear some similarity to those for individual atoms (chapter 3). If we imagine the two nuclei united together for a moment, then their electrons would take up the configuration appropriate for the atom with atomic number equal to the sum of those of the original nuclei. Separating the nuclei will produce an

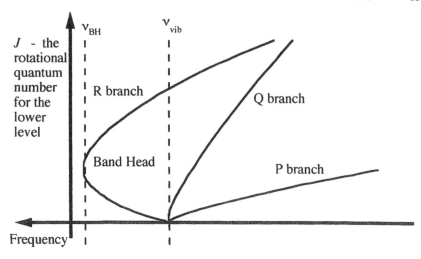

Figure 5.16 A Fortrat diagram: this is a plot of the frequencies of the rotational components of a vibrational transition against the rotation quantum number, J. The diagram shows the behaviour of the fine structure at much larger values of J than shown in figures 5.14 and 5.15. Note the reversal of the trend for the R branch leading to a concentration of lines at ν_{BH}, the Band Head. The shape of the R branch plot is that of a parabola, and the P branch plot, if reflected in the frequency axis, would be the continuation of that curve. On this diagram the Q branch, due to transitions with $\Delta J = 0$, is also shown for completeness, it produces a band head at ν_{vib}.

intense electric field along the axis between the nuclei, and the effect upon the electrons will be similar to that of an external field applied to the 'united atom'. As we have seen in chapters 2 and 3, the orientation of the angular momentum of both individual electrons and atoms as a whole is quantized in the presence of an external field. At high field strengths the coupling of L and S breaks down (the Paschen–Back effect) and their orientations are individually quantized. The total orbital angular momentum is then represented by the quantum number M_L which takes values $-L, (-L+1), (-L+2), (-L+3), \ldots, (L-1), L$. In an electric field, however, the energy level for $-M_L$ has the same energy as that for $+M_L$ (unlike the case in a magnetic field; see Zeeman and Stark effects). Differing values of $|M_L|$ though will have very different energies because of the strength of the field. Now L will be precessing very rapidly around the field direction (i.e. around the internuclear axis) and so will have little meaning as an angular momentum. The energy levels are thus more appropriately defined by the value of $|M_L|$.

For atoms, the value of L was indicated through the use of letters: S, P, D, F, G etc for $L = 0, 1, 2, 3, 4 \ldots$. A similar practice is used for diatomic molecules, except that the equivalent Greek letters are used. Thus the values of $|M_L|$ of $\Lambda = 0, 1, 2, 3, \ldots$ etc are indicated through the use of the symbols Σ,

Π, Δ, Φ, etc for the corresponding energy levels. The Σ level is single valued, but the Π, Δ, Φ, etc levels are doubly degenerate because M_L has values $\pm\Lambda$.

The spin angular momentum, S, is unaffected by an electric field, but does precess in the presence of a magnetic field. For levels with $\Lambda > 0$ (Π, Δ, Φ, Γ, etc levels), there is an internal magnetic field in the direction of the internuclear axis due to the orbital motions of the electrons. For such levels, the spin orientation is quantized. The values of M_S are symbolized by Σ (not to be confused with the Σ symbol for the $\Lambda = 0$ level) and take values from $-S$, $(-S + 1)$, $(-S + 2)$, $(-S + 3)$, ..., $(S - 1)$, S. The level is thus split into $(2S + 1)$ components, and as with atoms, this is called the multiplicity. It is added to the term symbol as the left superscript. For the Σ level there is no internal magnetic field and the S vector is fixed in space, nonetheless, the multiplicity is still used.

The angular momentum, denoted by Ω, is obtained by adding Λ and Σ, but as both are aligned to the internuclear axis, only simple addition is involved, not the vector addition needed for atoms to obtain J. So we get

$$\Omega = |\Lambda + \Sigma|. \tag{5.18}$$

This is added as a following subscript in the level symbol.

If the molecule is reflected in a plane passing through both nuclei, then the eigenfunction arising from the solution of the Schrödinger equation may either be unchanged or become negative. These two possibilities are indicated in the symbol for the Σ term by a following superscript $+$ or $-$ respectively.

Another commonly encountered notation is to prefix the symbol with a Roman letter. X is reserved for the ground state, and A, B, C, etc used for excited states of the same multiplicity. States with a different multiplicity are labelled in a similar way but using lower case letters.

Finally, for homonuclear diatomic molecules, the wavefunction may be even or odd (i.e. its sign may remain the same or reverse) with respect to the interchange of the nuclei. This is indicated by a following subscript; 'g' for even and 'u' for odd (from the German, *gerade* and *ungerade*).

Thus, as examples, we have the ground level of the hydrogen molecule as $^1\Sigma_g^+$ or $X^1\Sigma_g^+$, the first excited level of the hydrogen molecule as $^1\Sigma_u^+$ and $^3\Sigma_u^+$, the ground level of the CH molecule as $A\,^2\Pi$ and the $^3\Delta$ level split into the three states $^3\Delta_1$, $^3\Delta_2$ and $^3\Delta_3$.

5.4.2 Energy

The energies involved in the electronic levels are much greater than those for either vibrational or rotational levels. Electronic transitions therefore usually also involve vibrational and rotational changes as well. As before, however, we shall use the terminology 'electronic transition' here for this state of affairs and not the more cumbersome electronic–vibrational–rotational transition sometimes

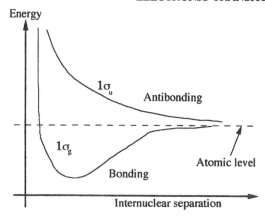

Figure 5.17 Variation of energy with internuclear separation for a diatomic molecule.

used elsewhere. The energy gaps are typically a few eV, and so the resulting spectral lines appear in the optical part of the spectrum (figure 5.2). There is no formula for electronic energy levels, and their actual values have to be found experimentally.

The electronic energy levels of a diatomic molecule must be intermediate between those of the 'united atom' and the two sets of atomic levels we would get for well separated individual atoms. If we start with the well separated atoms, and allow them to approach each other, then the original energy levels will split. The two new levels, in quantum mechanical terms, represent electron distributions in which there is either a high probability or a low probability of finding the electrons between the two nuclei in the finally formed molecule. The first possibility is symmetric with respect to interchange of the nuclei and is labelled with a subscript 'g' (for *gerade*, or even, see above). The second is antisymmetric and is labelled with a subscript 'u'. Thus the hydrogen molecule can have electrons in the $1\sigma_g$ or $1\sigma_u$ orbitals (equivalent to the 1s level in atomic hydrogen). The variation of energy with internuclear distance for the two cases is shown in figure 5.17. From that diagram we may see that only for the $1\sigma_g$ orbital will there be a stable hydrogen molecule. The two types of orbital are thus often called bonding and antibonding respectively. The hydrogen molecule is unstable with even one electron in an antibonding orbital, but other molecules may have such electrons. Thus the lithium molecule has a ground state given by

$$1\sigma_g^2 \quad 1\sigma_u^2 \quad 2\sigma_g^2.$$

Higher energy levels in the original atoms exhibit similar behaviour to that of the ground state as the molecule is formed.

Molecules more complex than the diatomic ones considered here can sometimes be dealt with in a manner similar to that for the diatomics. This applies especially when the molecule is linear, like CO_2, or has some symmetry, like H_2O. The details of these and of other more complex cases are beyond the

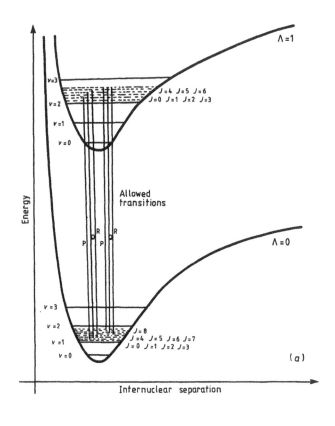

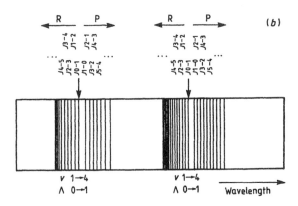

Figure 5.18 Electronic transitions in a diatomic molecule and the resulting spectrum. Reproduced from *Molecular Spectra and Molecular Structure. Vol 1 Spectra of Diatomic Molecules* by G Herzberg (New York: Van Nostrand Reinhold).

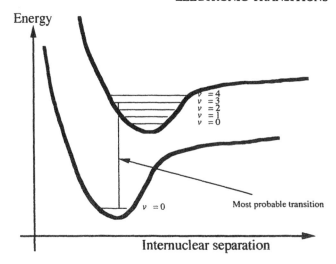

Figure 5.19 The Franck–Condon principle. The $v = 0$ to $v = 3$ transition is more probable than the $v = 0$ to $v = 0$ transition because no change in the internuclear separation is involved.

scope of this book and the interested reader is referred to the bibliography for sources of information on such molecules.

5.4.3 Selection rules

In diatomic molecules we have the following selection rules for normal electronic transitions:

$\Delta\Lambda = 0$ or ± 1

$\Delta\Sigma = 0$

$\Delta v = \pm n$ (where n is a small integer)

$\Delta J = \pm 1$

($\Delta J = 0$ is also allowed except within $\Lambda = 0$ to $\Lambda = 0$ transitions and for the $J = 0$ to $J = 0$ transition). Note the change to the constraint on vibrational changes from the pure vibrational case for which $\Delta v = \pm 1$ only.

5.4.4 Transitions

The energy change in an electronic transition is the sum of the changes in electronic, vibrational and rotational levels (equation (5.2)). The electronic transition by itself would normally produce a spectral line in the optical region (figure 5.2). The vibrational changes will split this line into a number of components typically separated from each other by a few nanometres. The rotational changes will then cause each of these components to be subdivided with separations of a few hundredths of a nanometre. At low dispersions, the rotational components will blend and result in the appearance of the molecular spectrum as a series of continuous bands (figure 5.18). Transitions involving the dissociation or formation of the molecule will also produce bands in the optical region, but these will be genuinely continuous. The rotational components will belong to P and R (and for some transitions to Q) branches, just as for the vibrational transitions. The band head can be to the high or low frequency end of the band depending on whether the internuclear separation is lower or higher in the upper level than in the lower level.

The speed of changes to the electron structure of a molecule is much faster than the speed of its vibrations. An electronic transition will therefore take place with the same internuclear separation for the initial and final levels. This results in some vibrational changes during an electronic transition being more probable than others even though all are allowed by the selection rules. This is especially the case when the mean internuclear separations differ between the lower and upper levels (the Franck–Condon principle) as shown in figure 5.19.

6

Radiation in the Presence of Fields

6.1 ZEEMAN EFFECT

6.1.1 Atoms

We have seen that the orientation of total angular momenta for both electrons and atoms is quantized (chapters 2 and 3). In the absence of an external field however, the $(2J + 1)$ states comprising an energy level are degenerate in the strict physical sense. That is, they have identical energies.

Such degeneracy may be removed if the particle is in the presence of a magnetic field. The transitions then occurring will produce photons of differing energies for the different pairs of states involved. In a spectrum, the lines will be observed to split into two, three or more components when the material producing them is in a magnetic field. This effect is called the Zeeman effect and applies to both emission and absorption lines, in the latter case strictly being called the inverse Zeeman effect. The selection rule applying to these transitions requires $\Delta M = 0, \pm 1$, except that $M = 0$ to $M = 0$ is forbidden for $\Delta J = 0$. Lines with $\Delta M = 0$ are usually identified with a π, those with $\Delta M = \pm 1$ using a σ.

A similar effect, the Stark effect, arises for electric fields. The latter, however, are rarely of significance in astrophysics since the very high conductivity of plasmas causes any charge differences within them to be smoothed out very quickly, but the effect is of importance for molecular spectra (chapter 5) and for line broadening (chapter 13) and is considered further below.

In the normal Zeeman effect, which applies to singlet ($S = 0$) lines, the spectral lines split into three components (figure 6.1). One of these is at the normal wavelength of the line, the others are shifted to longer and shorter wavelengths by an amount

$$\Delta\lambda = \frac{e}{4\pi m_e c^2} \lambda^2 g B \qquad (6.1)$$

where B is the field strength; g is called the Landé factor, and is a correction factor whose value is usually close to unity, and given by

$$g = 1 + \frac{J(J+1) + S(S+1) - L(L+1)}{2J(J+1)}. \qquad (6.2)$$

In the optical region, therefore, the splitting is by about 0.01 nm per tesla $(10^{-11} \text{ m T}^{-1})$.

For the situation sketched in figure 6.1, the line splits into three components, each being triply degenerate. The radiation absorbed or emitted for each component is polarized. When the line of sight is perpendicular to the magnetic field direction, all the components are linearly polarized, the directions for the outer (σ) components being perpendicular to the field and that for the central (π) component parallel to the field. When the line of sight is along the magnetic field direction, then only the two outer components are seen, and these are circularly or elliptically polarized in opposite directions. At other orientations, a mixture of the two limiting cases is observed. A qualitative understanding of how the polarizations arise can be obtained by imagining the projected motions that would be observed for a free electron in a magnetic field. Assuming that there was no motion along the field direction, then looking along the magnetic field, the electron would be seen to be moving in a circular path. Perpendicularly to the field, the motion would appear to be a linear harmonic oscillation across or along the line of sight (figure 6.2).

If the Landé factors for the two levels differ, then the energy degeneracy of the transitions is lifted. Each allowed transition can then produce a separate spectral line (figure 6.3). The resulting effect upon the spectrum in this case is called the anomalous Zeeman effect, since unlike the normal Zeeman effect, it cannot be explained by classical physics.

At very high field strengths (half a tesla or more) some of the components start to combine, a phenomenon known as the Paschen–Back effect. This arises because the L and S angular momenta decouple from each other and then couple individually to the magnetic field. At moderate spectral resolutions therefore the spectrum may look as though the normal Zeeman effect is operating, though each of the three components in such a case is actually a close multiplet.

At extreme field strengths ($> 10^3$ T), such as might be found on white dwarfs and neutron stars, the quadratic Zeeman effect becomes important. The lines are then all displaced to higher frequencies by an amount

$$\Delta\lambda = \frac{-\varepsilon_0 n^4 h^3}{32\pi^3 e^2 m_e^3 c^3} \lambda^2 (1 + M^2) B^2. \qquad (6.3)$$

6.1.2   Molecules

A molecule in a magnetic field has the degeneracy of its levels removed just as does an atom. Thus a rotating molecule may only take up such directions with

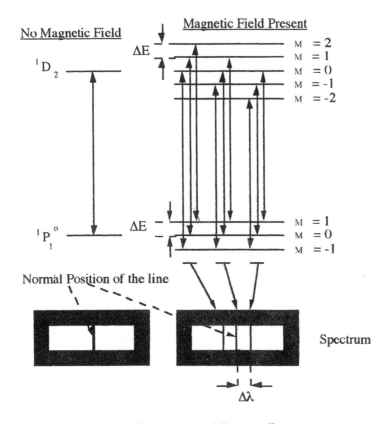

Figure 6.1 The normal Zeeman effect.

respect to the field that its angular momentum component in the field direction is quantized. A rotational level will then split into $(2J + 1)$ components. However, given the large number of components already present in molecular bands, the investigation of the Zeeman components is very difficult, even in the laboratory.

6.2 STARK EFFECT

6.2.1 Atoms

The Stark effect is a splitting of spectral lines observed when the emitting or absorbing material is in the presence of an electric field. The separation of the components is usually on the same scale as the fine structure of the line even for fields of 10^7 V m^{-1}. As already remarked, it has little direct relevance to astronomy because electric fields are rarely encountered. It does, however, have a bearing on molecular spectra (chapter 5) and on the pressure broadening of spectral lines (chapter 13).

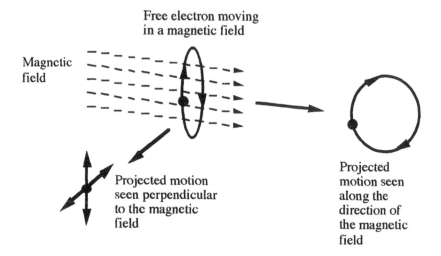

Figure 6.2   Components of the motion of a free electron in the presence of a magnetic field.

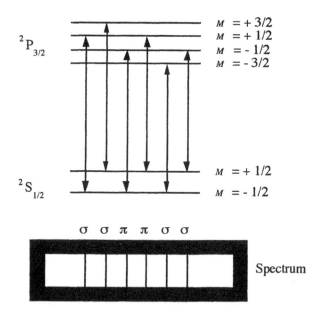

Figure 6.3   The anomalous Zeeman effect.

The manner of operation of the Stark effect is rather different from and less simple than that of the Zeeman effect. The electric field displaces the positive nucleus from the centre of the electron cloud. This displacement results in an electric dipole whose magnitude is proportional to the intensity of the external field. The dipole moment, however, depends not only upon the field strength but also upon the orientation of the atom to the field. Thus the dipole moment depends upon the orientation of the total angular momentum vector, J, to the field direction. J precesses about the field direction at a rate dependent upon the field strength, and its component in the field direction is quantized with values $M = -J, (-J + 1), (-J + 2), \dots, (J - 1), J$ as for the Zeeman effect.

Now the splitting of the energy levels induced by an electric field will depend upon the product of the dipole moment and the field strength. But since the dipole moment also depends upon the field strength, the splitting will vary with the square of that strength. The separation of the components of a spectral line will thus also depend upon the square of the field strength.

A second difference from the Zeeman effect is that the dipole moment induced by the field will not be altered by reversing the field direction. The energy shift is therefore identical for $+M$ and $-M$. The number of states into which an energy level splits under the influence of an electric field is thus $(J + 1/2)$ or $(J + 1)$ depending upon whether J is half-integral or integral. Except for $M = 0$, these states will be doubly degenerate. This contrasts with the $(2J + 1)$ non-degenerate states produced by a magnetic field.

There are no simple or general formulae for the energies of the individual states produced by the Stark effect. Thus for hydrogen for example, we have

$$\Delta E = 1.27 \times 10^{-24} n(n_2 - n_1)F + 1.03 \times 10^{-39} n^4$$
$$\times \left[17n^2 - 3(n_2 - n_1) - 9(n - n_2 - n_1 - 1)^2 + 19\right] F^2 \quad (6.4)$$

where ΔE is the energy change from the field-free level, n is the principal quantum number, n_1 and n_2 are electric quantum numbers which take integer values from 0 to $(n - 1)$ and F is the field strength in V m^{-1}. The resulting line splitting is complex: the Balmer hydrogen-α line for example has over 30 Stark components.

At very high field intensities the L and S vectors decouple, as with the Paschen–Back effect, and are then independently quantized to the field direction.

6.2.2 Molecules

The Stark effect for molecules again causes splitting of the energy levels. The dipole induced by the field causes second-order effects to occur even for zero dipole molecules. The effect is extensively used in experimental spectroscopy but is of little interest to the astrophysicist.

6.3 FREE–FREE RADIATION

6.3.1 Introduction

Classically, radiation is emitted or absorbed whenever a charge is accelerated,

$$\frac{\mathrm{d}E}{\mathrm{d}t} = -\frac{e^2 a^2}{6\pi \varepsilon_0 c^3}$$

(6.5)

where a is the acceleration of the charge and e is the magnitude of the charge (assumed equal to that of the electron). It was the impossibility of explaining the lack of such radiation from electrons in atoms that led to Quantum and Wave Mechanics (chapter 2). However, such radiation is found for free electrons being accelerated. In one form it leads to the Synchrotron and Gyrosynchrotron radiation discussed below. In another form it produces Free–Free radiation. This latter form of radiation occurs when an electron is accelerated by the electric field of a nearby ion. Theoretically the ion is also accelerated by the encounter, but its mass is so large in comparison with that of the electron that the effect is negligible.

Although the electron does not 'belong' to the ion, we may still represent the transition on an energy level diagram. It is a transition from one level above the ionization limit to another level also above the ionization limit (figure 6.4). The levels above the ionization limit represent relative translational motion between the electron and ion. Although translational motion is quantized, the levels are so close together that, except for conditions of extreme density such as may be found in white dwarfs and neutron stars, they effectively form a continuous band. A photon of any energy can therefore be emitted or absorbed in such a transition. The name for the radiation comes from the electron being free of (i.e. not bound to) the ion both before and after the transition.

If we think of the electron and ion in particle terms, then the electron would bypass the ion in a hyperbolic orbit if no radiation were lost or gained (figure 6.5). The effect of such energy loss or gain is to cause the electron's path to change to a different but still hyperbolic orbit about the ion. Much of the time we are concerned with the emission rather than the absorption of energy by the free–free process. The electron is thus decelerated with respect to the ion during the transition. An alternative name for the process is therefore Bremsstrahlung Radiation, from the German for braking.

In a single non-relativistic encounter between an electron and an ion, the emission spectrum is flat with an intensity of

$$I(v) = \frac{Z^2 e^6}{24\pi^4 \varepsilon_0^3 c^3 m_e^2 b^2 v^2}$$

(6.6)

where $I(v)$ is the emission per unit volume, Z is the atomic number of the ion (assumed completely ionized), v is the relative velocity of the electron and

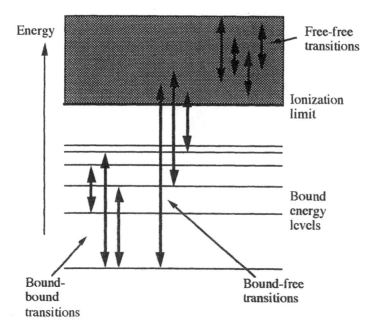

Figure 6.4 Free–free transitions on an energy level diagram.

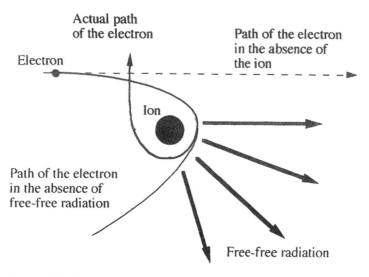

Figure 6.5 Free–free radiation as Bremsstrahlung (braking) energy loss.

ion, and b is the minimum separation of the ion and electron for the electron moving in a straight line in its original direction. This quantity is usually called

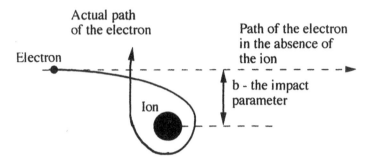

Figure 6.6 The definition of the impact parameter, b.

the Impact Parameter (figure 6.6). This is true up to a frequency of about

$$v \approx \frac{v}{b} \tag{6.7}$$

after which it tails off. Thus, for an encounter of an electron with a hydrogen ion (proton) at 10 km s^{-1} and with $b = 0.1$ nm, the emitted spectrum is flat from the longest radio waves to the near infrared, though of course the intensity from such a single event is small indeed.

6.3.2 Thermal free–free spectrum

If we have a hot plasma then the electrons and ions of which it is composed will be undergoing numerous close passes. The resulting spectrum, known as the Thermal Free-Free spectrum, must be found by integrating equation (6.6) over all possible values of b, and for the range of velocities of the particles in the plasma, given by Maxwell's distribution:

$$\frac{N(p)\,dp}{N} = \left(\frac{2}{\pi m^3 k^3}\right)^{1/2} \frac{p^2}{T^{3/2}} \exp(-p^2/2mkT)\,dp \tag{6.8}$$

where $N(p)$ is the number density of particles with momenta in the range p to $p + dp$, N is the total number density of the particles and m is the particle mass.

The details of this procedure are left for the interested reader to pursue from sources in the bibliography. Here, just the final result is quoted:

$$I(v) = \frac{e^6}{2^{1/2}3^{3/2}\pi^{3/2}\varepsilon_0^3 k^{1/2} c^3 m_e^{3/2}} \frac{g Z^2 N_e N_Z}{T^{1/2} \exp(hv/kT)} \tag{6.9}$$

$$= 6.79 \times 10^{-51} \frac{g Z^2 N_e N_Z}{T^{1/2} \exp(hv/kT)} \; \text{W}\,\text{m}^{-3}\,\text{Hz}^{-1} \tag{6.10}$$

where N_e is the electron number density, N_Z is the ion number density and g is a correction factor, known as the Gaunt Factor, whose value is a complex function of temperature and frequency. Usually its value lies between 0.2 and 5. See (for example) *Radiative Processes in Astrophysics* by G B Rybicki and A P Lightman, pp 160 and 161.

The total energy from a plasma due to free–free emission, neglecting absorption, may be found by integrating Equation 6.9 over all frequencies:

$$E(T) = \frac{k^{1/2}e^6}{2^{1/2}3^{3/2}\pi^{3/2}\varepsilon_0^3 hc^3 m_e^{3/2}} \bar{g}Z^2 N_e N_Z T^{1/2} \tag{6.11}$$

$$= 1.42 \times 10^{-40} \bar{g}Z^2 N_e N_Z T^{1/2} \quad \text{W m}^{-3} \tag{6.12}$$

where $\bar{g}$ is the Gaunt Factor for temperature T, averaged over frequency.

For a real plasma these last equations must also be integrated over all ionic states and over all the atomic species present. For a pure hydrogen plasma at a high temperature, however, we have $Z = 1$, $g \approx 1$, and $N_e = N_Z = N/2$, so that

$$E(T) \approx 1.42 \times 10^{-40} N_e^2 T^{1/2} \quad \text{W m}^{-3}. \tag{6.13}$$

In terms of the mass emission coefficient, ε_ν (equation (4.20)), we have

$$\varepsilon_\nu = \frac{e^6}{2^{1/2}3^{3/2}\pi^{3/2}\varepsilon_0^3 k^{1/2}c^3 m_e^{3/2}\rho} \frac{gZ^2 N_e N_Z}{T^{1/2}\exp(h\nu/kT)} \tag{6.14}$$

etc.

We may find the mass coefficient for free–free absorption, κ_ν, from the emission coefficient and Kirchhoff's law. We have encountered Kirchhoff's law previously (equations (1.1) and (4.41)) and it states that the ratio of emission and absorption coefficients is a constant. In the case of black body radiation, that constant has the value of the black body radiation intensity, and so from Planck's equation

$$\frac{\varepsilon_\nu(T)}{\kappa_\nu(T)} = \frac{2h\nu^3}{c^2(\exp(h\nu/kT) - 1)} \tag{6.15}$$

we get

$$\kappa_\nu = \frac{e^6}{6^{3/2}\pi^{3/2}\varepsilon_0^3 hk^{1/2}cm_e^{3/2}\rho} \frac{gZ^2 N_e N_Z(\exp(h\nu/kT) - 1)}{\nu^3 T^{1/2}\exp(h\nu/kT)}. \tag{6.16}$$

6.3.3   Interstellar nebulae

For many hot interstellar nebulae, such as planetary nebulae and H II regions, the plasma of which they are composed will be optically thick at low frequencies and optically thin at higher frequencies. The transition from one regime to the other occurs in the radio region, typically at frequencies of a GHz or so. We can therefore use the Rayleigh–Jeans approximation instead of the full Planck

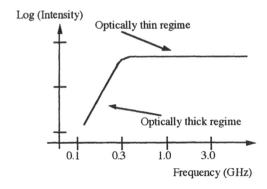

Figure 6.7   Schematic radio spectrum of a hot interstellar nebula.

formula to obtain κ_ν. The exponential factor in equation (6.14) will also be close to unity for all physically encountered situations. Hence we find for such nebulae that κ_ν varies as the inverse square of frequency.

The radio emission spectrum of a nebula will thus obey the Rayleigh–Jeans law in the optically thick regime, and its intensity will vary as the square of the frequency. In the optically thin regime, however, the observed intensity will also depend upon the optical depth. Now, the optical depth in turn depends upon κ_ν and so will also vary as the inverse square of frequency. Its product with intensity will thus be constant with frequency, and so the emitted spectrum will be flat (figure 6.7). The electron number density of the nebula can then be found from the flat part of its spectrum, and its temperature from the Rayleigh–Jeans portion.

6.3.4   Relativistic free–free spectrum

When the velocity between the ion and electron is relativistic, such as might be the case for cosmic rays, then the spectrum for an individual encounter is again flat, this time up to a frequency of about

$$\nu \approx \frac{(\gamma - 1)m_e c^2}{h} \tag{6.17}$$

where γ is the relativistic Lorentz factor:

$$\gamma = \frac{1}{\sqrt{1 - v^2/c^2}}. \tag{6.18}$$

The rate of energy loss is given by the Bethe–Heitler formula:

$$\frac{dE}{dt} = -\frac{e^6}{8\pi^2\varepsilon_0^3 hc^2 m_e}(\gamma - 1)Z(Z + 1.3)\left[\ln\left(\frac{183}{Z^{1/3}}\right) + \frac{1}{8}\right]N_Z. \tag{6.19}$$

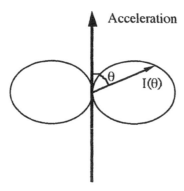

Figure 6.8   Cross-section (polar diagram) through the dipole radiation pattern from an accelerated charge. The three-dimensional shape must be obtained by rotating the polar diagram through 180° about the acceleration vector.

6.4   SYNCHROTRON AND GYROSYNCHROTRON RADIATION

6.4.1   Introduction

These forms of radiation are very closely related to free–free radiation and occur when a charged particle is accelerated by a magnetic field. The radiation is called Synchrotron if the particles are moving at relativistic velocities, and Gyrosynchrotron (or sometimes Cyclotron) when the velocities are lower. The terms Magnetic Bremsstrahlung and Magnetobremsstrahlung Radiation are also used. The particles involved are usually, but not always, electrons.

6.4.1.1   Gyrosynchrotron radiation.   We consider first gyrosynchrotron radiation. This is simpler though of less significance than synchrotron radiation for astrophysical applications, but it does occur in emissions from solar flares, white dwarfs and neutron stars.

The radiation from an accelerated, non-relativistic charged particle has a dipole pattern (figure 6.8) around its acceleration vector. The radiation is emitted at the gyro frequency, v'_G, and has intensity, $I(\theta)$, at an angle θ to the acceleration, given by

$$v'_G = \frac{eH}{2\pi m_e} \quad \text{and} \quad I(\theta) = \frac{e^2}{16\pi^2 \varepsilon_0 c^3} a^2 \sin^2 \theta \qquad (6.20)$$

where a is the acceleration and e is the charge (assumed to be that of the electron).

Consider an electron, since in practice electrons are the main particles producing gyrosynchrotron and synchrotron radiation in astrophysical situations, moving perpendicularly to a magnetic field. The moving electron constitutes an electric current in the magnetic field, and so by the Lorentz equation it experiences a force at right angles to both the field and the current (the left hand rule of school physics):

$$F = ev \times H \qquad (6.21)$$

where F is the force, v is the electron's velocity and H is the magnetic field strength. The electron therefore moves in a circular path whose plane is perpendicular to the field direction, and whose radius (the gyro radius) is given by

$$R_G = \frac{\gamma m_e v}{eH} \qquad (6.22)$$

where v is the component of the electron's velocity across the magnetic field. The acceleration is thus at right angles to the direction of motion of the electron, and the radiation, emitted instantaneously in a dipole pattern, becomes smoothed to isotropy over one complete 'orbit' of the electron.

6.4.1.2 Synchrotron radiation. The situation becomes more complex as the velocity of the electron increases to relativistic velocities, and the radiation changes from gyrosynchrotron to synchrotron. To an external observer the dipole radiation (figure 6.8) will be affected by the aberration due to the electron's motion. Radiation emitted by the electron at an angle α' to its direction of motion will appear therefore to the external observer to be at an angle α to the direction of motion. Using the relativistic aberration formula, since normally only high velocities are of significance for astrophysical situations, α is then given by

$$\alpha = \sin^{-1}\left(\frac{\sin \alpha'}{\gamma \left(1 + v \cos \alpha'/c\right)}\right). \qquad (6.23)$$

Thus the originally symmetrical emission around the acceleration vector will be concentrated into the forward direction. At 99.9% of the speed of light, for example, radiation originally emitted at 45° to the direction of the velocity, will be observed only 1° away from it. Similarly the nulls, at 90° in the electron's frame of reference, will appear at 2.5° in the external observer's frame of reference. Only radiation originally emitted between 177.4° and 180° to the direction of motion by such an electron will actually appear to be emitted in a rearward direction. The observed radiation from a charged relativistic particle being accelerated by a magnetic field is thus very tightly beamed around the particle's instantaneous forward direction (figure 6.9).

The original null points occur at an angle of 90° to the acceleration. In the plane of the velocity and acceleration vectors, this corresponds to $\alpha' = 90°$. Now since $v/c \approx 1$ and α is small, we find from equation (6.23)

$$\alpha \approx \gamma^{-1} \quad \text{radians.} \qquad (6.24)$$

Thus, to a good degree of approximation, the emitted radiation appears in a beam whose half width is $[1 - (v/c)^2]^{1/2}$ radians to an external observer.

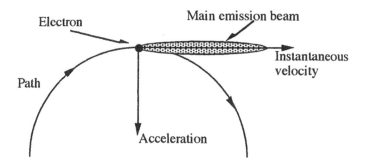

Figure 6.9   Cross-section through the synchrotron emission pattern.

The loss of energy causes the electron to spiral inwards to tighter and tighter orbits. The velocity and acceleration vectors are thus not truly at right angles to each other. The deviation from 90° is usually very small, however, and does not significantly affect the above analysis.

6.4.2   Synchrotron radiation spectrum

The radiation in the rest frame of the electron has an emission spectrum given by

$$I(v) \propto v^2 a(v)^2 \tag{6.25}$$

where $a(v)$ is the component of the acceleration of frequency v. The observed spectrum of the radiation from a relativistic electron in a magnetic field, however, is very different. An observer in the plane of an electron's motion will only be able to see its radiation while the main emission lobe is along his line of sight. From equation (6.24) therefore, this will occur for about $1/\gamma\pi$ of the 'orbital' period. The orbital period is given by the reciprocal of the relativistic gyro frequency

$$v_G = \frac{eH}{2\pi\gamma m_e} \tag{6.26}$$

so that radiation may potentially be seen for a time interval, Δt:

$$\Delta t = \frac{2m_e}{eH}. \tag{6.27}$$

However, relativistic particles will be moving only a little slower than their emitted radiation. As the electron emits the last visible photons, it will thus be close behind the first visible photons to be emitted. The observed pulse duration will thus be highly compressed in comparison with the prediction of equation (6.27). A photon emitted at the leading edge of the main emission beam will have travelled a distance $c\Delta t$ in the time that the electron travels $v\Delta t$. Photons from the trailing edge of the main emission beam are thus emitted

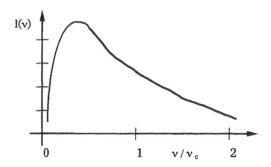

Figure 6.10 Schematic spectrum of the synchrotron emission from a single electron.

from a point $v \Delta t$ closer to the observer than those at the leading edge, and are a distance $(c - v) \Delta t$ behind them. The photons are hence received by the external observer only over a time interval

$$\Delta t_0 = \left(1 - \frac{v}{c} \right) \Delta t \approx \frac{\Delta t}{2\gamma^2} = \frac{m_e}{\gamma^2 e H} \qquad (6.28)$$

(using the approximation $(1 + v/c) \approx 2$). The reciprocal of this interval gives the maximum Fourier component of the received radiation

$$\nu_{\text{max}} = \frac{\gamma^2 e H}{m_e} = 2\pi \gamma^2 \nu_G' \qquad (6.29)$$

where ν_G' is the non-relativistic gyro frequency.

Thus the observed synchrotron spectrum consists of high harmonics of the electron's non-relativistic gyro frequency, up to the limit, ν_{max}. Overlap of the emission lines results in an apparent continuum spectrum, and a higher precision analysis gives for the envelope of this continuum

$$I(\nu) \approx 5 \times 10^{-25} (\nu/\nu_c)^{1/3} H_p \quad \text{W Hz}^{-1} \qquad (6.30)$$

for $\nu \ll \nu_c$, and

$$I(\nu) \approx 3 \times 10^{-25} (\nu/\nu_c)^{1/2} e^{-\nu/\nu_c} H_p \quad \text{W Hz}^{-1} \qquad (6.31)$$

for $\nu \gg \nu_c$, where H_p is the component of the magnetic field perpendicular to the electron's velocity vector and ν_c is a critical frequency which marks the approximate limit of significant emission, given by

$$\nu_c = 1.5 \gamma^2 \nu_G'. \qquad (6.32)$$

The maximum emission occurs at about $0.29 \nu_c$, and the half width of the emission is about $1.5 \nu_c$ (figure 6.10).

For the situation intermediate between gyrosynchrotron and synchrotron, the spectrum will contain the fundamental frequency, v'_G, and one or more harmonics. The number of harmonics will increase as v/c approaches 1 until the above synchrotron spectrum is reached.

Of course, in practice, synchrotron radiation comes from electrons with varying pitch angles to the magnetic field, and with a range of velocities, and quite probably the magnetic field will vary in its strength and direction over the emission region. If we take the commonly found exponential distribution for the electron energies,

$$N_e(v) \propto E^{-n} \qquad (6.33)$$

where $N_e(v)$ is the number density of electrons with velocities in the range v to $v + dv$ and E is the electron's kinetic energy, then this is much less sharply peaked than the emission from an individual electron, and we may usefully approximate an individual electron's emission as though its total energy were emitted over an interval, Δv_c, centred on v_c. The total emission averaged over all pitch angles for a single electron is proportional to γ^2, and for relativistic electrons we have

$$E = \frac{\gamma m_e c^2}{2}. \qquad (6.34)$$

Thus the flux from an electron of energy E, $F(E)$, is given by

$$F(E) \propto E^2 \qquad (6.35)$$

spread over the interval Δv_c. From equation (6.32) we have

$$v_c \propto \gamma^2 \propto E^2 \qquad (6.36)$$

and so

$$\Delta v_c \propto E \Delta E. \qquad (6.37)$$

The total flux at a given v_c is thus

$$F(v_c) = \frac{F(E)}{\Delta v_c} N_e(v) \Delta E \qquad (6.38)$$

$$\propto E^{1-n} \qquad (6.39)$$

or from equation (6.36)

$$F(v) \propto v^{(1-n)/2}. \qquad (6.40)$$

Thus the spectrum of a real synchrotron radiation source may be expected to follow a power law. The index in the form

$$F(v) \propto v^{-\alpha} \qquad (6.41)$$

is known as the spectral index, and we may easily see that it is related to the index of the electron energy distribution by

$$\alpha = \frac{n-1}{2}. \tag{6.42}$$

The value of the spectral index for astrophysical sources ranges between about 0.2 and 2.0. The spectrum of an astrophysical synchrotron radiation source thus shows an increase in intensity as the frequency decreases. In the radio region (where most but not all synchrotron emission is to be found) its spectrum is therefore in sharp contrast to that of thermal emission, for which the intensity follows the Rayleigh–Jeans law and decreases as the frequency decreases.

6.4.3 Limits to synchrotron emission

The synchrotron emission intensity of course cannot continue to increase indefinitely as the frequency decreases. Three effects conspire to limit its increase.

Firstly, imagine a disturbance that moves the electrons in a plasma. Since the ions are so much more massive than the electrons, they will remain essentially stationary, and a strong electric field will be produced tending to pull the electrons back towards their equilibrium positions. On moving back to these positions, they will overshoot and set up simple harmonic motion (ignoring energy losses). Thus there is a natural frequency at which the electrons as a whole in a plasma will oscillate. This frequency is called the critical or plasma frequency, ν_0, and is given by

$$\nu_0 = \frac{eN_e^{1/2}}{2\pi\varepsilon_0^{1/2}m_e^{1/2}}. \tag{6.43}$$

Any regular disturbance of the electrons in the plasma, including electromagnetic radiation, with a frequency near or less than the plasma frequency will be very strongly absorbed. This is shown by the behaviour of the refractive index of the plasma. In the absence of a magnetic field, this is given by

$$n(\nu) = \sqrt{1 - (\nu_0/\nu)^2} \tag{6.44}$$

and we may see that it reduces to zero at the plasma frequency, representing non-propagation of the radiation inside the plasma. Radiation incident onto such a plasma from the outside will be totally reflected at and below the plasma frequency. The plasma frequency is thus a cut-off below which no radiation can escape from the plasma.

The second effect in fact dominates the normal plasma cut-off and is called the Razin limit. The effect arises because a refractive index significantly less

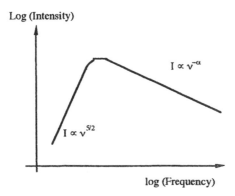

Log (Intensity)

$I \propto \nu^{-\alpha}$

$I \propto \nu^{5/2}$

log (Frequency)

Figure 6.11 Spectrum of a synchrotron radiation source at low frequencies showing self-absorption.

than unity represents a phase velocity for the wave significantly greater than c. The compression of the pulse due to the electrons being close behind previously emitted wavefronts (equation (6.28)) is then no longer significant because the electron velocity cannot equal or exceed c. There is thus a cut-off of synchrotron radiation for frequencies less than about $\gamma \nu_0$.

Finally, the material may become optically thick and synchrotron self-absorption will set in. For such an optically thick source, the spectrum is independent of the electrons' velocity distribution, and is simply proportional to the frequency to the power of 5/2 (figure 6.11).

6.4.4 Polarization

After the form of its spectrum, the other major characteristic of synchrotron radiation which distinguishes it from thermal emission is its polarization. This is generally elliptical polarization, becoming closer to linear polarization as the electron's energy increases.

Considering a single electron, its radiation is polarized with the electric vector parallel to the acceleration vector. Thus for electrons moving perpendicular to a magnetic field, the observed projection of the acceleration vector will be perpendicular to the magnetic field, and the observed radiation will be linearly polarized (figure 6.12). When the electron has a component of its velocity along the magnetic field, the emitted beam of radiation sweeps around a cone centred on the magnetic field direction (figure 6.13). As can be seen from the diagram, the plane of the linear polarization rotates in space as the emission beam sweeps around this velocity cone. A single observer sees the radiation only while his line of sight intersects the emission beam just as before, but now during that interval the plane of the polarization rotates through a small angle. In the example drawn in figure 6.13, the rotation for the observer is clockwise. Thus there is a small degree of circular polarization to be added to the linear polarization; i.e.

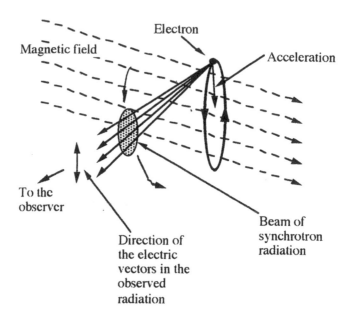

Figure 6.12 Polarization of synchrotron radiation perpendicular to the magnetic field.

the total radiation is elliptically polarized. However, the observer will also be able to see radiation from electrons whose velocity vectors do not pass directly through the line of sight, providing that they are within $1/\gamma$ radians of the line of sight. Across the emission beam, the polarization varies as shown in figure 6.14 accordingly as the observed direction of the projected acceleration of the electron changes. The direction of rotation of the plane of polarization will thus be modified by its variation across the emission beam. At high velocities, however, the effect of beams to one side of the line of sight will be cancelled by the beams on the other side, leaving just the basic effect. At low velocities such cancellation may not occur completely, leaving significant degrees of elliptical polarization in the observed (gyrosynchrotron) radiation.

For relativistic electrons, the emitted power falls rapidly as the angle to the perpendicular to the field direction increases. The observed radiation therefore originates from angles to the magnetic field near to 90°. The rotation of the plane of polarization by the sweeping of the emission beam around the velocity cone is then very small, and so the radiation is largely linearly polarized. For electrons with the velocity distribution given by equation (6.33) and in a uniform magnetic field, the observed degree of linear polarization, π_L, is given by

$$\pi_L = \frac{n+1}{n+7/3}. \tag{6.45}$$

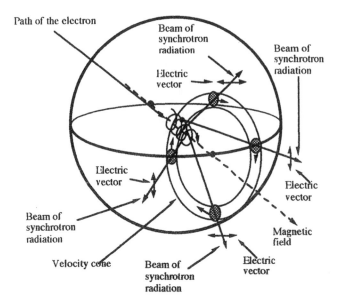

Figure 6.13 Polarization of synchrotron radiation from an electron with a non-zero velocity component along the magnetic field.

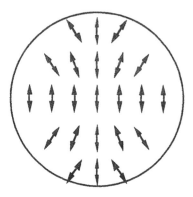

Figure 6.14 Schematic variation in the linear polarization across the synchrotron radiation emission beam.

For the range in the values of n from about 1.5 to 5.0 found for typical astrophysical synchrotron radiation sources, the observed degrees of linear polarization may thus be expected to range from 65% to 80%.

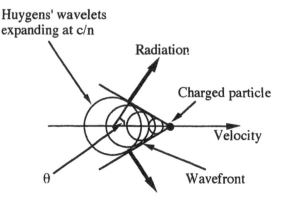

Figure 6.15   The Čerenkov radiation cone.

6.5   ČERENKOV RADIATION

Although not strictly belonging under the heading of radiation in the presence of fields, Čerenkov radiation is nonetheless most appropriately considered in this chapter. The radiation arises when a charged particle passes through a medium at greater than the local velocity of light. When the refractive index of a medium is greater than unity, the phase velocity of light is less than c. A relativistic particle can thus travel faster than that phase velocity. If we construct the Huygens' wavelets in such a situation (figure 6.15), then the envelope of the emission is a cone centred on the particle's direction of motion. Unsurprisingly, since both arise from the motion of an object through a medium at a velocity greater than that of the propagation of waves in the medium, the pattern is similar to that of the shock wave produced by a supersonic aircraft. In astrophysics the main importance of the radiation is for the detection of cosmic rays and neutrinos (via relativistic electrons produced by collision), and as a noise source in photomultiplier tubes and CCDs.

The angle of the emitted radiation to the velocity direction is

$$\theta = \tan^{-1}\left(\frac{n^2 v^2}{c^2} - 1\right)^{1/2} \qquad (6.46)$$

where n is the refractive index of the material, and its spectrum is given by

$$I(v) = \frac{e^2 v}{2\varepsilon_0 c^2}\left(1 - \frac{c^2}{n^2 v^2}\right). \qquad (6.47)$$

6.6   THE FARADAY EFFECT

Linearly polarized radiation has its plane of polarization rotated on passing through a magnetized plasma.   It is only of significance in astrophysical

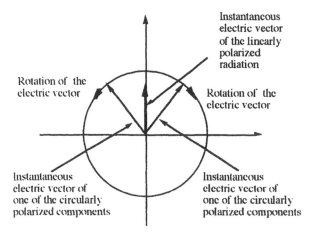

Figure 6.16   Linear polarization as the sum of two circularly polarized components.

situations for radio waves. The effect is called the Faraday effect and arises because linearly polarized radiation can be regarded as the sum of two circularly polarized components with opposite senses of rotation (figure 6.16). The effect of radiation passing through a plasma is to cause the charged particles (in practice only the electrons because of their low masses) to move. The presence of a magnetic field then constrains those motions to be spirals around the field direction. The two circularly polarized components of the radiation thus become realized in practice. For frequencies well above the gyro frequency (equation (6.26)), the refractive indices for the two components are

$$n(\nu\theta) = \sqrt{1 - \frac{(\nu_0/\nu)^2}{1 \pm (\nu_G' \cos\theta/\nu)}} \qquad (6.48)$$

where θ is the angle to the magnetic field. The plus sign in the denominator applies to one component and the minus to the other. The refractive indices for the two components thus differ by

$$\Delta n(\nu\theta) \approx \frac{\nu_0^2 \nu_G' \cos\theta}{\nu^3}. \qquad (6.49)$$

Since this depends upon the inverse cube of the frequency, the difference in the refractive indices becomes negligible outside the radio region. The different refractive indices mean that the phase velocities of the two circularly polarized components differ. The phase relationship between them will therefore change as the radiation propagates through the medium. This corresponds to a rotation of the direction of the linear polarized resultant (figure 6.17). The angle of the rotation in radians, for propagation through the interstellar medium is given by

$$\Psi \approx 8100\,\lambda^2 \int_0^D N_e B\,\mathrm{d}x \qquad (6.50)$$

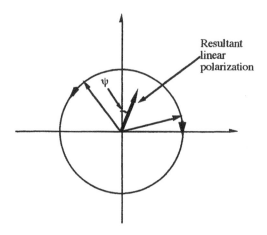

Figure 6.17   Rotation of the plane of linear polarization through a phase change in its circularly polarized components (cf figure 6.16).

where D is the distance travelled through the interstellar medium in parsecs, B is the component of the magnetic field along the line of sight in tesla, λ is in metres and N_e per cubic metre. When ψ is positive the magnetic field component along the line of sight is directed towards the Earth; if negative, then it is away from the Earth.

6.6.1 Dispersion

The refractive index of a material at radio frequencies is frequency dependent (equation 6.49), and the phase velocity therefore also varies with frequency. Radiation of mixed frequencies originating at the same moment will therefore be spread out in arrival times after passing through such a material.   For most astrophysical sources, the differential delay thus introduced into signals by the interstellar medium is not of much significance. The pulses from pulsars, however, are so sharp that the delay can be measured as a function of frequency, and is called the dispersion. When combined with measurement of the Faraday rotation, this can provide information on both the galactic magnetic field and the electron density of the interstellar medium.

7

Spectroscopy of Solid Materials

The spectral properties of solid materials, which include for our purposes those of liquids and even of dense gases, are not generally of great significance for the astrophysicist. They are of importance, however, for solar system studies and for consideration of the properties of the dust in the interstellar medium.

A simple point about solid spectroscopy that ought to be obvious but frequently is not is that it almost always refers to the surface layers of the material. These may be characteristic of the material as a whole, but equally often may be quite dissimilar in nature. Pure diamond for example is composed of carbon in its bulk properties. Its surface, however, has a layer of hydrogen atoms which terminate the free carbon bonds. On the Earth many materials have a surface layer of oxide which is uncharacteristic of the material as a whole. Spectroscopic studies of such materials may therefore be quite misleading with regard to their true properties.

Molecules within solids and other condensed phases are affected in several ways by the presence of the numerous nearby atoms, ions and other molecules. Firstly, the intermolecular forces within a crystal may cause a distortion of the structure of the molecule from that which it would take up in a free state, thus changing the energies of the various levels etc, and altering the wavelengths of the resulting features in the spectrum. Secondly, the interactions with other particles during transitions results in a broadening of the spectrum lines. The lines from rotational transitions and in many cases vibrational transitions, therefore blend into each other, and only very broad absorption or emission bands are observed (figure 7.1). This broadening is also of importance for lines from the electronic transitions of all particles, even in quite rarefied gases, and it has a significant effect upon stellar spectra, where it is known as Pressure Broadening (chapter 13). Thirdly, free rotation of the molecules is prevented or hindered, so that rotational features disappear even without the blending effect. Finally, for all types of particles (ions, atoms and molecules), the narrow electronic energy levels that we have encountered for the particles in their free state become broadened into wide pseudo-continuous allowed energy ranges called Bands. The lower energy bands, which would be occupied by electrons in the ground state, are called the Valence Bands, the higher ones are called the

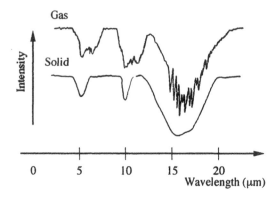

Figure 7.1 Schematic example of the difference between the spectrum of a molecule in the gaseous phase and that in a solid.

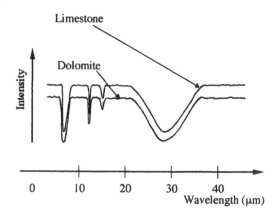

Figure 7.2 Infrared spectra of limestone ($CaCO_3$) and dolomite ($CaMg(CO_3)_2$).

Conduction Bands. These bands are involved in the properties of the material as an electrical conductor or insulator, and the interested reader is referred to books on solid state physics for further information. For our purposes their significance lies in the further broadening of spectral features to have widths in energy terms equal to the sum of the widths of the two bands involved in the transition.

In the laboratory, materials can be made pure enough for some useful information to come from solid spectroscopy. Outside the laboratory, solid spectra contain such little detail that dissimilar materials can have indistinguishable spectra (figure 7.2). The spectra can also vary markedly for slight changes in conditions or trace elements etc. For remote sensing of the Earth, the general nature of the material being observed may be known, or it may be possible to take samples from parts of the region being imaged (Ground Truth Data). Thus remote sensing can provide data on the extent

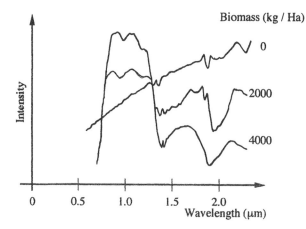

Figure 7.3 Example of the variation of solid spectra of vegetation with increasing biomass.

of ice in the polar regions, crop types and biomass densities, surface and atmospheric temperatures, some mineral deposits, cloud type and extent, water vapour pressure, etc (figure 7.3).

For astronomical objects other than the Earth, ground truth data cannot be obtained. On Mars and Venus, where landers have sampled the surface directly, some data are available, but these are often ambiguous. Only for our Moon do we have analyses comparable in detail to those available for the Earth. Even then, only a very limited number of sites on the Moon were sampled, and so the data are not necessarily widely applicable. Other information may give a clue to the nature of the surfaces of planets and satellites to aid in interpretation of solid spectra. Thus surface features analogous to those on the Earth may be observed. For example, the shield volcanoes on Mars and Venus suggest a basaltic type composition for the rocks. In other cases material from the surface may be vaporized and become available for study via atomic and molecular spectroscopy. Thus carbon compounds may be expected on the surfaces of comet nuclei, and sulphur compounds on the surface of Io (figure 7.4).

Outside the solar system, not even such limited guidance may be found to assist the interpretation of solid spectra. Dust grains are widely to be found in the interstellar medium, in dense molecular clouds and star-forming regions, and in the atmospheres of some cool stars. Some indication of a possible composition for the dust grains comes from the depletion of the interstellar medium in some elements, with those elements being presumed to form the grains. Thus very broad features in the spectrum of the interstellar medium at 200 nm, 3, 10 and 30 μm, are attributed, but not without ambiguity, to graphite, ices, silicates and MgS respectively. Numerous diffuse absorption features are found in the optical spectrum of the interstellar medium. The mains ones are at 443, 578, 618 and 628 nm, with widths of several nanometres. Their strength seems to show a

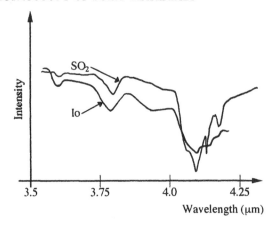

Figure 7.4 Laboratory spectrum of SO$_2$ compared with that of Io.

correlation with the number of dust grains, and so they are presumably produced by the grains, but at present their origin is not understood (chapter 16).

Part 2

Astronomical Spectroscopic Techniques

8

Optical Spectroscopes

8.1 INTRODUCTION

Detectors operating in the optical part of the spectrum are generally broad-band devices, sensitive to radiation over a wide range of wavelengths. The spectrum cannot therefore be observed directly as it may at shorter wavelengths where the detector's response varies with the energy of the photon. In the optical region a separate device is required to separate out the radiation into its component wavelengths before it is detected. That device is called a Spectroscope†.

There are several processes whereby the radiation can be separated into its component wavelengths. The simplest is to use a range of filters placed before the detector to isolate particular regions of the spectrum. Traditionally, however, this method of studying the spectrum of an astronomical object is regarded as a separate subject from spectroscopy, and is called Photometry. The spectral resolution, R, is defined as

$$R = \frac{\lambda}{\Delta\lambda} \tag{8.1}$$

where λ is the operating wavelength and $\Delta\lambda$ is the smallest wavelength interval that may be resolved. With this definition, photometry is concerned with spectral resolutions of less than about 100, spectroscopy with spectral resolutions greater than about 100. This is an arbitrary division, however, and exceptions occur. Thus studying the sun using an Hα filter is generally regarded as a branch of photometry even though the spectral resolution may be around 10 000, while identifying quasars on an objective prism plate (see later) at a spectral resolution of around 50 would belong to spectroscopy. Here the separation is therefore made by regarding spectral studies employing only filters not to be a part of spectroscopy.

† Strictly this term should be reserved for visual devices, and Spectrograph used when photography is used to detect the spectrum. Spectrometer should then be used for devices using other types of detectors such as CCDs. I prefer to use the single term, spectroscope, for these devices irrespective of the detector they employ on the grounds that this is the practice applied to the term Telescope, and we do not call it a Telegraph when a camera is attached in place of an eyepiece!

A second method of obtaining a spectrum is based upon differential refraction and results in rainbows (chapter 1) and prism-based spectroscopes. The latter are still used in astronomy and are briefly considered later in this chapter. Most modern astronomical spectroscopes, however, utilize interference effects in some way to produce the spectrum. The resulting devices which have found application in astronomy are Diffraction Gratings, Fabry–Pérot Etalons and Fourier Transform Spectroscopes, and we consider each of these below.

8.2 DIFFRACTION GRATINGS

The effect of a diffraction grating is nowadays familiar to most people from the colours reflected off compact discs.   These, like diffraction gratings, have numerous closely spaced parallel apertures, and the colours result from constructive and destructive interference between light reflected from those different apertures. Although a grating will usually contain many thousands of apertures, the principle of its operation is based upon the interference effects produced by just two apertures. Most gratings in practice are reflection gratings, but the principle of their operation is the same as that of a transmission grating. Since with the latter it is easier to visualize what is happening, the grating equation is derived for a transmission grating.

Let us consider then the Fraunhofer (far field) interference effects from a pair of closely spaced apertures (figure 8.1) illuminated by a distant monochromatic source of wavelength, λ. The path difference at an angle, θ to the incoming ray is given by (figure 8.2)

$$\Delta p = s \sin \theta \qquad (8.2)$$

where s is the separation of the apertures.

The interference pattern on the screen will be a series of light and dark fringes (figure 8.3b) accordingly as the path difference is an integer, or integer plus half, number of wavelengths and we have constructive or destructive interference, respectively. The intensities of the maxima are modulated by the interference pattern from a single aperture (figure 8.3a). The centres of the bright fringes correspond to integer wavelength path differences, and so at such points we have

$$\Delta p = s \sin \theta = n\lambda \qquad (8.3)$$

where n is an integer, usually referred to as the order of the interference. The angle between successive maxima is thus

$$\Delta\theta = \sin^{-1}\left(\frac{(n+1)\lambda}{s}\right) - \sin^{-1}\left(\frac{n\lambda}{s}\right) \qquad (8.4)$$

which for small angles simplifies to

$$\Delta p \approx \frac{\lambda}{s}. \qquad (8.5)$$

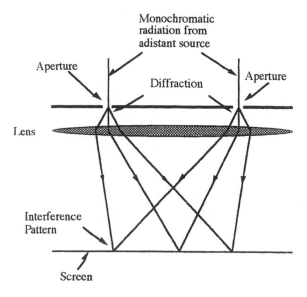

Figure 8.1 Fraunhofer interference from a pair of apertures.

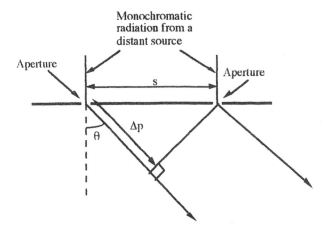

Figure 8.2 Path difference in Fraunhofer interference at two apertures.

If the normal to the plane of the apertures is at an angle f to the incoming radiation; corresponding to the often utilized inclined grating, then (figure 8.4)

$$\delta p = s \sin\theta + s \sin\phi = n\lambda \qquad (8.6)$$

We then get the grating equation in its normal form

$$\theta = \sin^{-1}\left(\frac{n\lambda}{s} - \sin\phi\right) \qquad (8.7)$$

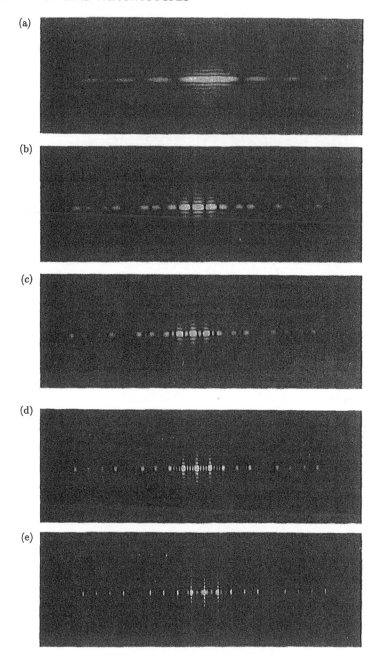

Figure 8.3 Fraunhofer interference patterns: (a) from a single aperture, (b) from two apertures, (c) from three equally spaced apertures, (d) from five equally spaced apertures, (e) from ten equally spaced apertures.

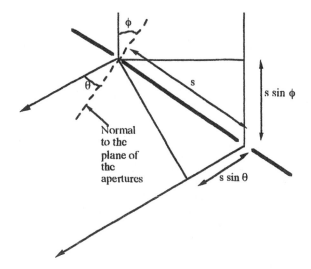

Figure 8.4 Plane of the apertures inclined to the incoming radiation.

From equation (8.7), we may see that the angular positions of the maxima are wavelength dependent. Maxima for different wavelengths will therefore appear at different angles to the incoming radiation (figure 8.5), and thus we have the principle of operation of the diffraction grating. Of course, as may also be seen from figure 8.5, the device at this stage is not much use as a practical means of producing a spectrum because the fringes are very broad and overlap. However if we add a third aperture in line with the first two and at the same spacing, then the fringe width reduces (figure 8.3c), while the separation remains constant. Weak secondary maxima also appear between the main fringes. Adding further apertures causes the primary fringe widths to reduce further (figure 8.3d), and although more secondary maxima are produced, they become progressively weaker. By the time the number of apertures is into double figures (figure 8.3e), the secondary maxima have almost disappeared and the primary maxima have become narrow enough for different wavelengths to be separated out (figure 8.6). In practice, gratings contain thousands of apertures and the angular fringe widths become equivalent to fractions of a nanometre.

8.2.1   Angular resolution

The above, almost qualitative, description of the operating principles of a diffraction grating may be made more quantitative via the intensity equation. The intensity in the interference pattern at an angle, θ, to the direction of the incoming radiation, $I(\theta)$, relative to the central intensity of the pattern, $I(0)$, is given by

$$\frac{I(\theta)}{I(0)} = \frac{\sin^2(\pi d \sin\theta/\lambda)}{(\pi d \sin\theta/\lambda)^2} \times \frac{\sin^2(N\pi s \sin\theta/\lambda)}{\sin^2(\pi s \sin\theta/\lambda)} \qquad (8.8)$$

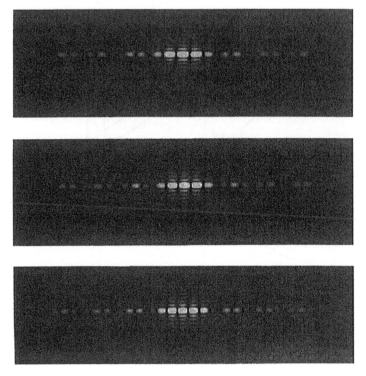

Figure 8.5 Interference patterns for monochromatic sources of different wavelengths through the same pair of apertures ($\lambda_1 > \lambda_2 > \lambda_3$).

where d is the width of an aperture and N is the number of apertures. The second fraction on the right hand side of equation (8.8) gives the widths and positions of the main fringes, and is often called the interference term. The first fraction represents a modulation of the interference pattern by the pattern for a single aperture (see figures 8.3a and 8.3e for example). Since N is an integer larger than unity, both numerator and denominator in the interference term tend to zero as ($s \sin \theta$) tends to an integer multiple of the wavelength. The interference term as a whole, however, tends to a limit of $\pm N$. Such positions correspond to the primary maxima, as we have previously seen (equation (8.3)).

The minima on either side of a primary maximum occur as the numerator of the interference term goes to zero. Thus we have for a primary maximum of order n, occurring at an angular position of

$$\theta = \sin^{-1}\left(\frac{n\lambda}{s}\right) \qquad (8.9)$$

the first zeros occurring at angular positions of

$$\theta \pm \Delta\theta = \sin^{-1}\left(\frac{(Nn \pm 1)\lambda}{Ns}\right). \qquad (8.10)$$

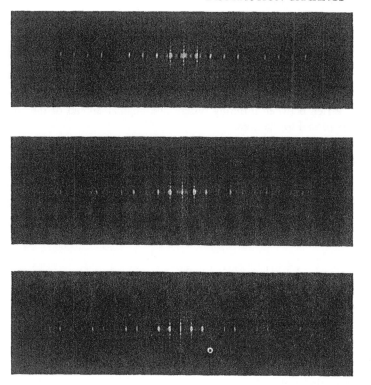

Figure 8.6 Interference patterns for monochromatic sources of different wavelengths through ten apertures ($\lambda_1 > \lambda_2 > \lambda_3$).

The angular width of a primary maximum is thus $2\Delta\theta$ and is given by

$$W = 2\Delta\theta \approx \frac{2\lambda}{Ns\cos\theta}. \tag{8.11}$$

The Rayleigh criterion for resolution is for one fringe maximum to be superimposed on the first zero of a second fringe pattern. Thus the angular Rayleigh resolution of a diffraction grating is given by $\Delta\theta$.

$$\text{Angular Rayleigh resolution} = \Delta\theta = \frac{\lambda}{Ns\cos\theta}. \tag{8.12}$$

8.2.2 Dispersion

The dispersion of a spectrum is the rate of change of wavelength with angular position. From equation (8.3) we have

$$\lambda = \frac{s\sin\theta}{n} \tag{8.13}$$

and so

$$\frac{d\lambda}{d\theta} = \frac{s \cos \theta}{n} \qquad (8.14)$$

Since θ is generally small, the dispersion along a spectrum from a diffraction grating is approximately constant ($\cos \theta \approx 1$ for small θ). The dispersion, however, varies between differing orders, the angular spread of the spectrum being proportional to the order.

8.2.3 Spectral resolution

We have defined the spectral resolution previously (equation (8.1)), and can now calculate it for a diffraction grating. From equations (8.12) and (8.14) we have

$$\frac{\lambda}{\Delta\lambda} = R = nN \qquad (8.15)$$

so that the spectral resolution of a grating spectrum depends only on the order of the spectrum and the number of apertures in the grating. For a typical astronomical spectroscope as used on a large telescope, N might be in the region of 10 000 to 20 000. The order of the spectrum would be 1 or 2, or occasionally 3. The typical spectral resolution available would therefore be in the region of 30 000, or about 0.02 nm in the visible part of the spectrum. For an echelle grating (see later in this chapter), used at order 100, the spectral resolution could reach 10^5 or more.

8.2.4 Free spectral range

The overlap between wavelengths shown in figure 8.5 is reduced to usable levels by the increase in the number of apertures in the grating. However for a real spectrum, with wavelengths extending from the radio region through to the far ultraviolet and beyond, there will still be an overlap between differing orders. The difference in wavelength between two superimposed wavelengths from adjacent orders is called the Free Spectral Range, and is usually given the symbol, Σ. If λ_1 and λ_2 are two such superimposed wavelengths, then from equation (8.9) and for small q, we have

$$\Sigma = (\lambda_1 - \lambda_2) \approx \lambda_2/n. \qquad (8.16)$$

For spectroscopes operating at low spectral orders, therefore, the free spectral range is large, and any overlap can be eliminated by the use of filters. At large orders, however, Σ can reduce to a few nanometres. The overlapping orders then have to be separated by a second spectroscope whose dispersion is perpendicular to that of the first. This arrangement is called a Cross Disperser, and it is discussed further along with echelle gratings and Fabry–Pérot etalons later in this chapter.

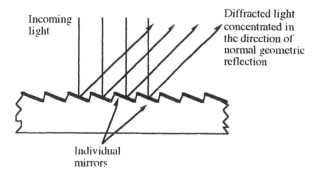

Figure 8.7   Enlarged section through a blazed reflection grating.

8.2.5   Blazing

For simplicity, we have derived the equations for diffraction gratings using apertures. The use of apertures in a real device would result in a transmission grating. In practice most gratings use mirrors in place of the apertures and result in reflection gratings. This causes no change to any of the above analysis. There are two reasons for the general preference for reflection over transmission gratings. The first is that reflection gratings are easier, and therefore cheaper, to produce (see below). The second is that reflection gratings are easier to blaze.

A major disadvantage of diffraction gratings for the production of spectra is obvious from figure 8.3e, and is that the available light is split into ten or more different orders. The spectrum that is being observed will therefore contain 10% or less of the incoming light. Since the overall intensities of the maxima are modulated by the diffraction pattern for a single aperture (figure 8.3), we may concentrate the light into a single order by widening the apertures until the central fringe of that modulation is just wider than a single spectrum. With simple apertures, however, this would concentrate the light into the zero order, and this, as we may see from equation (8.14) is not a spectrum. The solution, for reflection gratings is to tilt the individual mirrors (figure 8.7) so that the single aperture maximum is directed towards the order of spectrum that is of interest. This is called 'Blazing the grating', and can concentrate up to 90% of the incoming light into the desired spectrum. Transmission gratings can also be blazed, by replacing the aperture by a prism to refract the light into the desired spectral order.

Blazing has the further advantage of reducing the problem of overlap of different spectral orders (see above). But as may be seen from figure 8.7, it will cause some loss of light through the vertical portions of the grooves intercepting part of the reflected beam. This problem is called shadowing, and can become significant if the reflection is at a large angle to the normal to the grating.

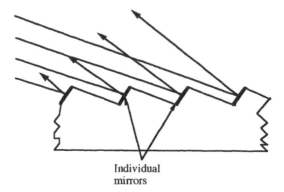

Individual
mirrors

Figure 8.3 Enlarged section through an echelle grating.

8.2.6 Curved gratings

All the gratings so far considered have been plane; they must therefore have other optical components to collimate and focus the light (see later in this chapter). Those other components will cause light loss and scattering, and in the case of lenses may introduce some chromatic aberration. This can be avoided by forming the grating on a curved (usually spherical) surface. The grating can then act as its own collimating and focusing components. For laboratory use spectroscope designs are often based upon the Rowland circle. This has a diameter equal to the radius of curvature of the grating, and the entrance slit, grating and focused spectrum all lie on the circle. For astronomical use such designs are usually too cumbersome, and more compact designs such as the Wadsworth (see later in this chapter) are preferred.

8.2.7 Echelle gratings

From equation (8.15) we may see that spectral resolution depends upon the order of the spectrum as well as the number of grating lines. An alternative to increasing the number of lines in order to improve resolution is therefore to observe the high order spectra. With a conventional grating, the high orders are at large angles to the normal to the grating, and shadowing by the vertical parts of the grating would obscure them completely. With an echelle grating, this is avoided and spectra with orders from several times ten to several times a hundred can be used. The echelle grating has the blaze angle greatly increased, and it is illuminated close to normal to the groove surfaces, and therefore at a large angle to the grating normal (figure 8.8). The coarser such a grating is made the higher order it can operate at, for as may be seen from the diagram, the path difference between adjacent light beams is about twice the depth of the groove. For laboratory work, grooves may be a tenth of a millimetre deep, but for astronomical use somewhat finer grooves would be more normal.

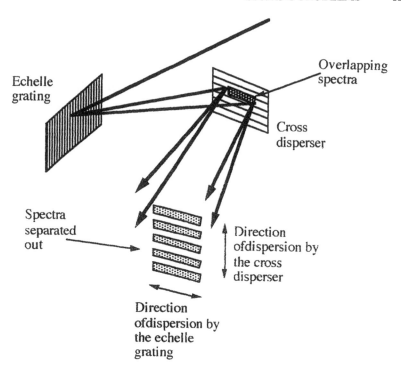

Figure 8.9   Schematic view of an echelle grating and a cross disperser.

The free spectral range (equation (8.16)) at high orders is very small, and the coarseness of the grating means that the individual apertures are quite large. The envelope of the diffraction pattern for a single aperture is therefore narrow. An individual order will thus cover only a small wavelength range before the modulation by the aperture diffraction pattern reduces its intensity to zero. It will then be overlapped by a different short wavelength interval from the next order spectrum. The final image thus consists of many short spectra superimposed onto each other. These are separated out by a second disperser acting perpendicularly to the direction of dispersion of the echelle (figure 8.9). This second disperser is called a Cross Disperser, and it can be of much lower dispersion than the echelle. The final spectrum then consists of an array of the short spectra (figure 8.10). Careful design of the system is needed to ensure that the short spectra overlap slightly in wavelength terms so that the complete spectrum can be synthesized.

8.3   GRATING PROBLEMS

We have already seen that the efficiency of a grating is reduced because it shares the available light amongst several orders, and that blazing can improve that efficiency, but not back to 100%. A second problem is that of overlap of adjacent orders, and this has to be solved by filters if the free spectral range is large, or by a cross disperser if it is small. So far, however, we have assumed that the grating itself is perfect. That is to say that the individual rulings are all identical, equally spaced and parallel. In practice, any real grating will fall short of perfection, and its deficiencies will reduce the quality of the final spectrum.

The effects of grating deficiencies may most easily be envisaged via the spectroscope's instrumental profile, or point-spread function. The instrumental profile is the response of the instrument to an ideal monochromatic source. A perfect spectroscope would produce the spectrum of such a source as an emission line at a single wavelength. An optically perfect, but real spectroscope would have that emission line spread over a small range of wavelengths through diffraction effects at the various apertures within the instrument (figures 8.3 and 8.11), and this spread constitutes the instrumental profile. The width of the instrumental profile determines the spectral resolution (equation (8.15)), and so reduces as the number of grating lines increases and/or the spectral order increases.

The shape of the instrumental profile may be calculated from the Fourier Transform† of the grating. The perfect spectroscope can then be seen to require an infinite number of apertures, and the transmission or reflectance to vary sinusoidally across each aperture. The instrumental profile of the optically perfect but real spectroscope then has the shape of the Fourier transform for a regularly spaced finite set of delta functions, modulated by the Fourier transform for a single aperture with 100% transmission or reflectance across it (figure 8.3e).

It is easy to see how errors in the production of the grating will affect the instrumental profile, by considering the effect on the Fourier transform. The errors are of three main types. Random errors in the spacings and/or parallelism of the rulings will correspond to the function to be transformed containing

---

† Fourier Transforms are a mathematical means of splitting a function into its component frequencies. The Inverse Fourier Transform then recombines the frequencies back into the original function. One version of their mathematical expression is

Fourier transform:

$$F(s) = F(f(x)) = \int_{-\infty}^{\infty} f(x)\, e^{-2\pi i x s}\, dx \qquad (8.17)$$

Inverse Fourier transform:

$$f(x) = F^{-1}(F(s)) = \int_{-\infty}^{\infty} F(s)\, e^{2\pi i x s}\, ds. \qquad (8.18)$$

In the usual terminology, $f(x)$ is the function in the spatial domain, and its Fourier transform, $F(s)$, is the function in the frequency domain. The modulus of the Fourier transform is called the power spectrum, and is a plot of the relative contributions to the original function at each frequency. The shape of the power spectrum corresponds to the shape of the instrumental profile.

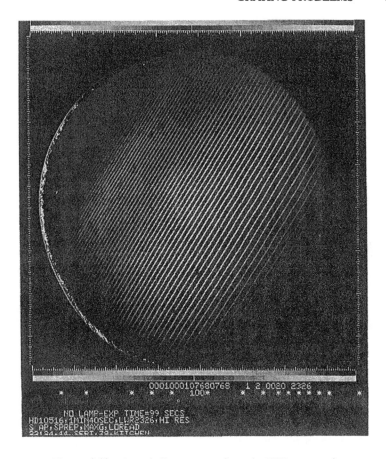

Figure 8.10   An echelle spectrum from the IUE spacecraft.

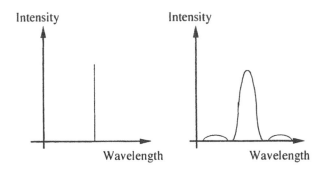

Figure 8.11   Instrumental profiles for a perfect spectroscope (left) and an optically perfect but real spectroscope (right).

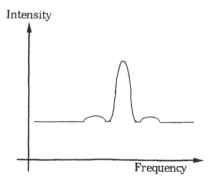

Figure 8.12 Power spectrum (i.e. instrumental profile) for a diffraction grating with random errors in the line spacing.

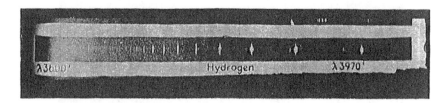

Figure 8.13 Spectrum showing Rowland ghosts due to a single periodic error in the grating rulings. Reproduced from *Fundamentals of Optics* (1976) by F Jenkins and H White with the permission of McGraw-Hill, Inc., New York.

white noise. The power spectrum of white noise is flat, and this will then have the transform for a regular grating superimposed (figure 8.12). Thus random errors will produce a background illumination, reducing the contrast in the final spectrum, but not affecting spectral resolution unless the errors become very gross. If the spacing and/or parallelism of the rulings vary in a periodic or multiply periodic manner, then the power spectrum will contain components corresponding to the frequencies and harmonics of those periodic errors. These will show up in the final spectrum as spurious spectral lines, known as Ghosts. A single periodic error produces Rowland ghost lines (figure 8.13) symmetrically about the true line and close to it.

Two or more incommensurate periodic errors produce Lyman ghosts which can occur at very different frequencies from the true line. Finally, there may be a progressive error in the rulings, so that their spacing increases or decreases in a regular manner. This has the effect of defocusing the collimated beam of light, so that it diverges or converges. It can largely be corrected by adjusting the focusing of the other optical components of the spectroscope.

Somewhat similar to ghosts, but not due to any deficiency of the grating, are the Wood's anomalies. These are sudden brightenings of the spectrum with a sharp onset and a slightly less sharp decline to longer wavelengths. The anomalies arise from that energy that would go into spectral orders behind the grating, if these were possible, being redistributed back into the actually visible orders. The additional light is almost 100% plane-polarized perpendicular to the grating rulings, and so the existence of such anomalies can be checked through the use of a polaroid sheet.

8.3.1   Production of gratings

Gratings are originally made by drawing a diamond across an aluminium or magnesium blank. These metals are chosen because they are soft and so reduce the wear on the diamond. The shape of the grooves (figure 8.7) is obtained by shaping the point of the diamond. The machines that are used to rule gratings vary in their design, but typically the blank will be moved along one axis while the diamond moves at right angles to that motion. The position and rate of motion of both the blank and the diamond are monitored and controlled interferometrically. Positional accuracies of 3 nm and parallelism to 0.005 seconds of arc are achievable. Spectral resolution (equation (8.15)) requires gratings to have as many lines as possible. Wear of the diamond point, which leads to a change in the groove profile, limits the number of lines possible to a few times 10^5, depending upon the length of the grooves. The wear of the diamond is also affected by the loading that is placed on it. Thus many more lines can be ruled if the blaze angle is small than if it is large because the indentation required in the surface of the blank is reduced. Similarly, echelle gratings (above), requiring coarse deep grooves are limited in the number of grooves that can be produced.

Original gratings are rarely used directly. Instead, replicas of the original are made. A thin layer of plastic or resin is applied to the original, and stripped off after setting. The replica is then mounted onto a rigid substrate and aluminized. If a transmission grating is required then the replica is mounted onto a transparent substrate and not aluminized. Many replicas can be made from a single original, so reducing the cost of the grating.

A completely different approach to the production of gratings results in so-called Holographic Gratings. These have a sinusoidal form to the grooves, but are effectively unlimited in size and number of lines. They are produced by illuminating a layer of photoresist coated onto an optical flat with two lasers to produce straight line fringes. Etching the photoresist then leaves the grating on the optical flat.

8.4   DIFFRACTION GRATING SPECTROSCOPES

In order to produce a usable spectrum, the diffraction grating must usually be combined with several other components, and it is the combination that we refer to as a Spectroscope. Most designs of spectroscopes incorporate in one form

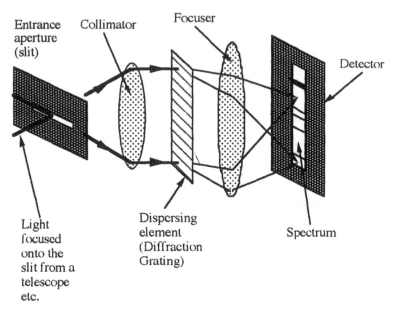

Figure 8.14   Basic components of a spectroscope.

or another the following basic components (figure 8.14): an Entrance Slit to reduce overlap between adjacent wavelengths and to reduce background noise, a Collimator to produce a parallel beam of light, a Disperser (in this case a diffraction grating), a Focusing Element to produce a focused image of the spectrum, and finally a Detector.

8.4.1 Resolution

A monochromatic source, as we have seen, will have an observed spectrum in which the light is spread to a greater or lesser extent into the neighbouring wavelength's positions. For a normal source with light of many wavelengths, the spectrum will therefore consist of an infinite number of such instrumental profiles, one for each wavelength present in the original spectrum, all overlapping each other. The spectral resolution (equation (8.15)) is set by the point at which the central peak of one instrumental profile is centred on the first dark fringe in the instrumental profile of a nearby wavelength. Such resolution can only be achieved in practice if it is not degraded by other components of the spectroscope. The collimator and focuser will normally be at least as large as the dispersing element. Their angular Rayleigh resolutions are given by

$$\alpha = \frac{1.22\lambda}{D} \qquad \text{radians} \qquad (8.19)$$

where D is the diameter of the lens or mirror (assumed circular) in the same units as the wavelength (λ). The diffraction-limited resolutions of the collimator and focuser should not therefore degrade the spectral resolution of the disperser. However, their optical quality needs to be sufficiently adequate to ensure that their aberrations also do not significantly affect the image.

Generally of more significance for the spectral resolution is the width of the entrance slit. Let us consider the geometrical optics of the basic spectroscope (figure 8.14): since the beams for individual wavelengths exiting a good quality plane diffraction grating will still be collimated, the image of the slit will have a width, w, at the detector given by

$$w = \frac{f_f}{f_c} W \qquad (8.20)$$

where W is the actual width of the slit, f_f is the focal length of the focuser and f_c is the focal length of the collimator. Clearly, the image of the slit must have a width no larger than that of the diffraction-limited instrumental profile for the remainder of the spectroscope, if it is not to degrade the spectral resolution. Thus from equations (8.14) and (8.15), we have

$$W < \frac{N f_c \lambda}{s \cos \theta} \qquad (8.21)$$

as a constraint on the width of the entrance slit for it not to degrade the spectral resolution. For practical astronomical spectroscopes, this constraint often means that the slit width must be less than the size of a stellar image at the focus of the telescope. Therefore either some light must be wasted, or an image dissector (see below) used, or the slit width increased and some spectral resolution sacrificed.

A further limitation on spectral resolution is the size of the individual detecting elements (pixels) of the detector. These must be no larger than the physical size of the core of the instrumental profile when projected onto the detector, if resolution is to be preserved. For an optimum design of spectrometer, we should have all four constraints on spectral resolution equal to each other, i.e.

Spectral resolution of the disperser
    = projected slit width
    = optical resolution of all the components of the spectroscope
    = detector pixel size.

Frequently astronomical spectroscopes have to take other constraints into account, such as that the physical size and weight of the spectroscope must be such that it can be mounted on the telescope, the focal ratio of the collimator must match that of the telescope, the focuser must be of short focal ratio to give reasonable exposure lengths, and by no means least, that the cost must remain within budget. All practical spectroscopes (chapter 10) therefore end up as compromises.

8.4.2 Throughput

Exposures to obtain the spectrum of an object with a given signal to noise ratio clearly need to be much longer than those to image the object directly to the same signal to noise ratio, because the available light is spread out into the spectrum. In practice, however, the required exposure is likely to be significantly longer than would be expected just on the basis of the greater area covered by the spectrum. This is because the spectroscope is not 100% efficient. If the entrance slit is smaller than the size of the stellar image at the telescope focus and an image dissector is not used, then the efficiency can fall to 10% or less. Spectroscopic exposures are then hundreds or thousands of times longer than those for direct images.

The efficiency of a spectroscope is measured by its Throughput, u (the terms Etendu and Light Gathering Power are also used). The throughput is defined as the amount of energy passing into the final spectrum when the entrance aperture is illuminated by unit intensity per unit area per unit solid angle. The throughput is given for spectroscopes with entrance slits by

$$u = A\tau\Omega \qquad (8.22)$$

where A is the area of the aperture, τ is the fractional transmission of the optics of the spectroscope and Ω is the solid angle of the slit as seen from the collimator, i.e.

$$\Omega \approx A/f_c^2. \qquad (8.23)$$

For slitless spectroscopes (see below), A is the area of the collimator, or focuser, or effective area of the dispersing element, whichever is the smaller, while Ω is the solid angle actually accepted by the spectroscope–telescope combination.

Unless only a very small spectral range is required, any transmission optics cannot have anti-reflection blooming applied because this is wavelength-dependent. There will thus be a light loss of at least 5% at each transmission surface. Mirror surfaces can be re-aluminized regularly, and so losses kept to 5% to 10% at each reflection. The widely used reflection grating, however, cannot be re-aluminized, and so losses due just to reflection efficiency could be from 10% to 20%. Shadowing, and energy going into unwanted orders could add another 10% to 50% to these losses. Thus for a Wadsworth spectroscope (see below) with an efficient grating, we might expect τ to be up to 0.7. But for a spectroscope using the not uncommon design of an achromat for the collimator, and a Schmidt camera for the focuser, with an older or less efficient grating, τ could fall to 0.3 or less. Losses due to the stellar image being larger than the slit are additional to these due to the optics and are not included in the figure for the throughput.

For a typical spectroscope carried at the Cassegrain focus of a major telescope, we might have a slit one or two hundred microns wide, and a few millimetres long, and f_c from 0.1 m to 0.5 m. The throughput would thus be in the range 10^{-9} to 10^{-14} for SI units. For a spectroscope at the Coudé or Nasmyth focus of a telescope, the collimator could have a focal length up to 10 m, resulting in throughputs down to 10^{-16}.

A useful parameter for comparing the overall performances of spectroscopes is the product of spectral resolution and throughput

$$P = Ru. \qquad (8.24)$$

Since spectroscopes for use at the (fixed) Coudé focus usually have much greater spectral resolutions than those at the (moving) Cassegrain focus of a telescope, their overall performances become comparable. Other things being equal, P will generally be highest for Fabry–Pérot etalon-based spectroscopes (see below), and lowest for prism-based spectroscopes, with diffraction gratings coming between.

8.4.2.1 Limiting magnitude. The faintest object from which a useful spectrum may be obtained determines the limiting magnitude of the spectroscope–telescope combination. Its value will clearly depend upon the throughput, but also upon many other factors. Thus the detector to be used makes a considerable difference, with CCDs having upwards of a hundred times the quantum efficiency of photographic emulsions. With most detectors, however, sensitivity varies with wavelength, so that basic CCDs and other silicon-based detectors become only a few times better than photographic emulsions in the blue part of the spectrum. With some detecting systems, such as the IPCS (Image Photon Counting System), noise reduction techniques permit given signal to noise ratios to be reached more rapidly than with other detectors of equal quantum efficiency. Secondly it will depend upon the purpose for which the spectrum is required, and the signal to noise ratio needed. Thirdly it will depend upon the nature of the spectrum; a nebula can have its spectrum found with surprising ease in comparison with that for a star since the light from the former is concentrated into a few strong emission lines with little or no continuum emission. The commonly used integrated magnitude for extended objects such as nebulae and galaxies will, by contrast, be far too optimistic since only a small fraction of the object will be covered by the spectroscope's entrance slit.

Finally, but by no means least, limiting magnitude will depend upon the patience of the observer and the generosity of the panel awarding the telescope time! Spectroscopic exposures generally have a low background and the observer can therefore continue to integrate the signal for long periods. Hubble, when he obtained the spectra of galaxies, used exposures extending over several nights. While such long exposures are unlikely today, since CCD exposures are limited by cosmic ray noise, the equivalent exposure can be built up by summing several spectra from shorter exposures after the cosmic ray spikes have been removed.

Clearly, no precise value for the limiting spectroscopic magnitude is possible. However, as a guide for comparing different spectroscope–telescope combinations, which is useful when the star's image is larger than the slit and it is trailed along the slit to widen the spectrum, we have Bowen's formula:

$$m_{\text{limit}} = 12 + 2.5 \log \left(\frac{W D_c D_t g Q t (d\lambda/d\theta)}{f_c f_f \alpha H} \right) \tag{8.25}$$

where m_{limit} is the faintest B magnitude giving a usable spectrum, D_c is the diameter of the collimator and D_t is the diameter of the telescope's objective. When the slit is wider than the star's image, the exposure will vary as D_t^2. g is the optical efficiency of the whole spectroscope–telescope combination. It therefore includes τ, and typically has values in the range 0.1 to 0.3. Q is the fractional quantum efficiency of the detector, t is the exposure time in seconds, α is the angular size of the star's image, typically 5 to 20 μrad (1″ to 4″), and H is the height of the spectrum.

For extended objects the situation is much more complicated, and different limiting magnitudes will be found for different types of object. The exposures required will generally need to be found by experiment. For a given object, however, it is useful to remember that the exposure for a given signal to noise ratio will vary as the square of the focal ratio of the telescope.

8.4.3   Designs of diffraction grating spectroscopes

The basic spectroscope shown in figure 8.14 can be used as illustrated, or used with a reflection grating in place of the transmission grating and/or with one or both of the collimator and focuser replaced by reflection optics. Reflection optics and grating are generally to be preferred over a transmission grating and refractive optics because of the intrinsic achromatism of reflection. Most astronomical spectroscopes employ plane diffraction gratings in either the basic design or Littrow (figure 8.15) or Ebert (figure 8.16) designs.

Saving on components and therefore also on light losses and cost can be made by the use of a curved reflection grating. As previously mentioned, designs based upon the Rowland circle are used for laboratory spectroscopes (e.g. figure 8.17). Normally, however, these designs are too cumbersome to be suitable for astronomical use. Not only are they physically large and heavy and therefore difficult to mount on a telescope, but their size makes it impossible to avoid differential flexure as the gravitational loading on the instrument changes with the telescope's attitude. The precise alignment required for the components of the spectroscope will thus be ruined. A commonly encountered design is therefore the Wadsworth (figure 8.18). At the price of an extra reflection, this gives a much more compact instrument.

Echelle gratings can be incorporated into designs similar to those above, but with the additional need, normally, for a cross disperser. One neat design, based upon the Littrow, simply places a prism immediately in front of the echelle grating (figure 8.19).

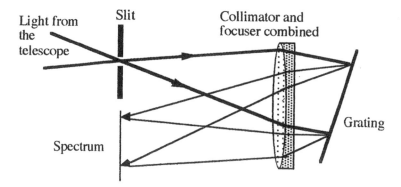

Figure 8.15 The Littrow design for a grating spectroscope.

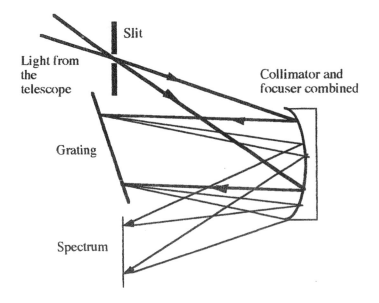

Figure 8.16 The Ebert design for a grating spectroscope.

For extended sources, a much longer slit than normal may be used for the spectroscope. The spectrum is then not trailed and so a particular horizontal section of the spectrum corresponds to a particular part of the image (figure 8.20). The spectra of many points across the image can thus be acquired in one exposure. Long slit spectroscopes do not differ in principle from the designs discussed above, but more consideration needs to be given to the performance of the instrument away from its optical axis. In particular, astigmatism and curvature of the spectral lines may become important.

All designs so far considered employ a slit as the entrance aperture to the

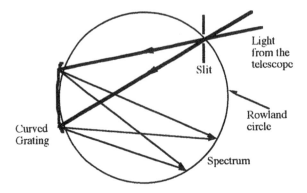

Figure 8.17 Design for a grating spectroscope based upon a curved diffraction grating and the Rowland circle.

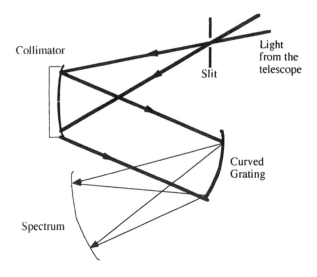

Figure 8.18 The Wadsworth design for a grating spectroscope.

spectroscope. While this has many advantages, it can also result in significant light losses. For some purposes, the slit can be dispensed with, resulting in a much more efficient system. Slitless spectroscopes come in two classes. The first has the disperser placed before the telescope objective. Normally a prism is used for this purpose since all the light is then concentrated into one spectrum. Such objective prism spectroscopes are considered later in this chapter. A transmission grating could be used in a similar fashion, but since it would be almost impossible to blaze, it would be very inefficient. Very coarse objective gratings are, however, sometimes employed in astrometry, since it can be easier to measure the positions of the first-order spectra which appear as faint star-

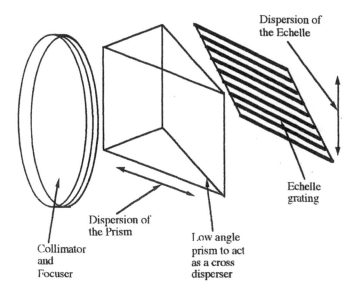

Dispersion of
the Echelle

Echelle
grating

Dispersion of
the Prism

Low angle
prism to act
as a cross
disperser

Collimator
and
Focuser

Figure 8.19 One design for a cross disperser for use with an echelle diffraction grating in a Littrow-type spectroscope.

like images on either side of the zero-order image than that of the overexposed zero-order image.

The other class of slitless spectroscopes contains standard spectroscope designs without the entrance slit, or with a slit larger than the size of the image at the focal plane of the telescope. For stellar images, this would normally degrade the final spectrum into near uselessness. It is therefore something of a paradox that such spectroscopes are actually used for extended objects. Slitless spectroscopes can be used when the spectrum of the object has little or no continuum, but is simply composed of emission lines. Such spectra occur for many objects including planetary nebulae, H II regions, supernova remnants, the solar chromosphere etc (chapters 14 and 16 and figure 1.4). The final spectrum then consists of a series of monochromatic images of the object in the light of each of its emission lines.

Grating spectroscopes operating outside the visual region (370–700 nm), but still within the optical region (100 nm to 20 μm) do not differ much in their optical principles from the standard designs discussed so far. Their differences lie in matching the details of the design to the requirements for the new wavelengths. Thus clearly the detector to be used must be able to respond to the wavelengths forming the spectrum. In the ultraviolet, the constraints on the optical components are more stringent because of the $\lambda/8$ requirement on mirror surface accuracies if the Rayleigh resolution is not to be degraded. The reflectivity of aluminium decreases with wavelength, and is down to 70% at 200 nm, so other surface coatings may need to be used. If transmission

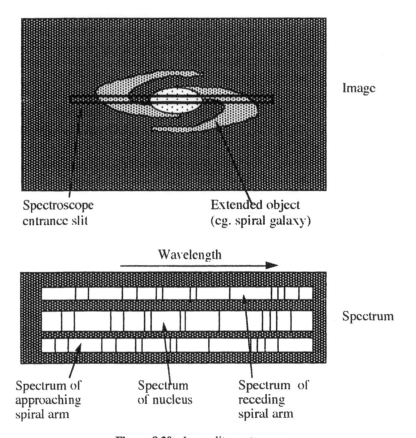

Figure 8.20 Long slit spectroscopy.

optics are used, then they must be fabricated from materials such as fused quartz or rock salt because glass absorbs at short wavelengths. Not least, for an instrument operating at wavelengths shorter than about 350 nm, it must be flown on a rocket, spacecraft, balloon or high altitude aircraft to get it above some or all of the Earth's atmosphere, which would otherwise absorb the radiation. In the infrared, similar considerations apply. The increased wavelength, however, reduces the requirements on surface accuracy for the optical components. At the longer infrared wavelengths, it may be necessary to cool the grating or even the whole instrument in order to reduce noise levels. Materials such as germanium, fluorite or proustite need to be used for transmission components. Although the atmosphere does not absorb completely over the infrared region, it will still be highly advantageous to use the infrared spectroscope from a high altitude observatory or to lift the instrument even higher by balloon etc.

Examples of specific designs for spectroscopes in current use are discussed in chapter 10.

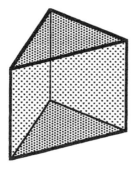

Figure 8.21   A prism as used in spectroscopy.

8.5   PRISMS

The classical and earliest spectroscope design employs a prism (chapter 1), and is the type most likely to have been encountered at school etc. Strictly the term 'prism' denotes an object with a constant cross section in one axis. In spectroscopic and common usage, however, it refers only to such an object with a triangular cross section (figure 8.21). It would be unusual to design an astronomical spectroscope today which used a prism as the primary disperser. This is because the prism needs to be large to have a reasonable spectral resolution. A large prism would be expensive, heavy, and would absorb the radiation passing through it. Moreover, comparable spectral resolution could be achieved with a quite cheap diffraction grating. Older spectroscopes which do employ prisms are still in use, and a prism may be used as a cross disperser (see echelle gratings above and Fabry–Pérot etalons below) or as an objective prism (see below). A brief discussion of their properties is therefore included here.

A prism acts as a disperser through the effect of differential refraction. That is to say, the refractive index of the material from which it is composed varies with wavelength, and so rays of differing wavelengths are deviated to different extents on passage through the prism (figure 8.22). From the law of refraction (Snell's law):

$$\frac{\sin i}{\sin r} = \mu \tag{8.26}$$

where μ is the refractive index, i is the angle of incidence of the ray to the normal to the surface and r is the angle of refraction of the ray to the normal to the surface, we may easily derive the equation for the deviation of a ray of particular wavelength (and therefore for a particular value of μ):

$$\theta = i - \alpha + \sin^{-1}\left\{\mu \sin\left[\alpha - \sin^{-1}\left(\sin i/\mu\right)\right]\right\} \tag{8.27}$$

where θ is the deviation (figure 8.22) and i is the angle of incidence to the normal to the first surface of the prism.

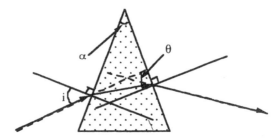

Figure 8.22   Deviation of a light ray by a prism.

Table 8.1   Refractive indices of some optical materials.

Substance	Wavelength (μm)						
	0.2	0.4	0.6	0.8	1.0	1.5	2.0
Crown glass		1.530	1.516	1.510	1.506	1.500	
Dense flint glass		1.652	1.618	1.608	1.603		
Fused quartz	1.600	1.470	1.457	1.452			
Fluorite	1.496	1.442	1.431	1.430	1.429	1.427	1.425
Rock salt	1.890	1.570	1.544	1.537	1.532	1.528	1.525

For the often utilized situation with the ray passing through the prism symmetrically (i.e. parallel to the base—the condition for minimum deviation and highest image quality), the equation simplifies to

$$\theta = 2\sin^{-1}\left[\mu\sin(\alpha/2)\right] - \alpha \tag{8.28}$$

and simplifies even further for the normally used apex angle of 60° to

$$\theta = 2\sin^{-1}(\mu/2) - 60°. \tag{8.29}$$

Typical values for the refractive indices of some relevant optical materials are listed in table 8.1.

Inspection of equation (8.28) and table 8.1 shows that, since the change in refractive index with wavelength decreases with increasingly wavelength for all the optical materials, a prism-based spectroscope will become increasingly ineffective at longer wavelengths. Thus even though rock salt and fluorite transmit in the infrared, they would not generally form useful dispersers for infrared spectroscopes.

The change in refractive index with wavelength may be described by the empirical Hartmann formula:

$$\mu_\lambda \approx A + \frac{B}{\lambda - C} \tag{8.30}$$

Table 8.2   Hartmann constants.

	A	B	C
Crown glass	1.500	3.5×10^{-8}	-2.5×10^{-7}
Dense flint glass	1.650	2.1×10^{-8}	1.5×10^{-7}
Fluorite	1.429	5.3×10^{-10}	3.6×10^{-7}

where A, B and C are the Hartmann constants for a particular material. Typical values are listed in table 8.2.

The dispersion of a 60° prism is then given for glass-like materials by

$$\frac{d\theta}{d\lambda} \approx \frac{-57.3AB}{(\lambda - C)^2} \quad °\,m^{-1}. \tag{8.31}$$

The spectral resolution is determined by the spread function due to diffraction at the apertures represented by the entry and exit faces of the prism. For symmetrical passage of the light through the prism, these apertures are equal and produce a rectangular light beam with a width of

$$D = L\left[1 - \mu^2 \sin^2(\alpha/2)\right]^{1/2} \tag{8.32}$$

where L is the length of the face of the prism. The Rayleigh resolution for such a rectangular beam is λ/D, and so we find that the spectral resolution is given by

$$R \approx \frac{ABL\left[1 - 0.25\left(A + \frac{B}{\lambda - C}\right)^2\right]^{1/2}}{(\lambda - C)^2}. \tag{8.33}$$

For a reasonably typical prism for use in an astronomical spectrograph made of dense flint and with a side length of 0.1 m, this therefore gives a spectral resolution of around 15 000, or less than half that of an equivalent diffraction grating.

8.6   PRISM SPECTROSCOPES

A slit-type prism spectroscope design can be produced for most of the diffraction grating instruments discussed earlier by replacing the grating with the prism. The most commonly encountered such designs are the basic spectroscope (figure 8.14) and the Littrow spectroscope (figure 8.15). For the Littrow design, a 30° prism would be used, with the rear surface aluminized. The light is therefore reflected and passes twice through the prism making it equivalent to a single 60° prism. The reflected beam is returned nearly along the same direction as the incoming beam, and so passes through the system in a similar manner to

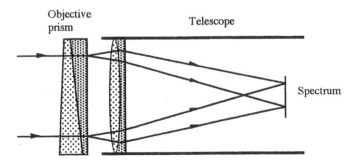

Figure 8.23   An objective prism spectroscope.

that shown in figure 8.15. There is no equivalent to the curved gratings, so the Rowland circle and Wadsworth designs have no prism-based equivalents.

A prism can have a diffraction grating applied to one of its surfaces, and the combination is then known as a Grism. In one form it provides a direct vision spectroscope for solar work, with the prism's deviation being equal and opposite to that of the blazed order of the transmission grating on its surface. The light then has no net deviation and is only dispersed, which simplifies the design of the whole instrument by eliminating off-axis components.

The simplest spectroscope of all is the objective prism. In this design, a thin prism covers the whole of the objective of the telescope, and so every object in an image is replaced by its spectrum. A simple prism may be used, in which case the telescope has to be pointed at a large angle to the desired direction because of the deviation caused by the prism. Alternatively two prisms of different glasses may be used in opposition to give zero deviation while still retaining some dispersion (figure 8.23). This is the exact inverse of the achromatic lens, where lenses of two different glasses are combined to give (close to) zero dispersion while still retaining focusing (i.e. deviation). The objective prism is a very efficient design because not only does it dispense with the slit and so improve the throughput hugely (see earlier discussion), but many spectra can be obtained in a single exposure. If an objective prism is applied to a Schmidt camera, then upwards of 10^5 spectra can be obtained on a single plate.

Unfortunately, the objective prism spectroscope also has serious disadvantages. The most important is the difficulty of finding radial velocities from the spectra because there is no comparison spectrum (chapter 9). Various attempts to solve this difficulty have been made, of which the most successful is to take two exposures on the same plate, with a slight displacement of the telescope and with the prism reversed between the exposures. Each object in the image then has two adjacent spectra (figure 8.24). Radial motions of the objects will give rise to Doppler shifts, causing the separation of a particular line in one spectrum from the same line in the second spectrum to change. If some of the objects have radial velocities known already, then that change in separation can be calibrated to give the velocities of all the objects on the plate. Other

Spectrum from
the first exposure

Spectrum from the second
exposure with the prism
reversed

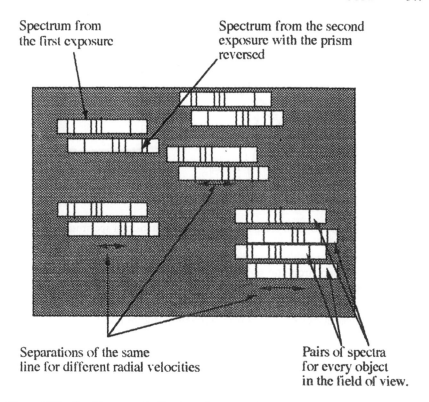

Separations of the same
line for different radial velocities

Pairs of spectra
for every object
in the field of view.

Figure 8.24 Doubly exposed objective prism plate showing the change in the separation of identical lines in pairs of spectra with the changing radial velocities of the objects.

drawbacks include the low dispersion of the spectra, and the weight and cost of the prism for a large telescope.

8.7 FOURIER TRANSFORM SPECTROSCOPE (MICHELSON INTERFEROMETER)

The interferometer used by Michelson and Morley (figure 8.25) in their classic experiment of 1887 to try to detect the Earth's motion through the aether can form the basis of a type of spectroscope known as a Fourier Transform Spectroscope. Unlike the diffraction grating or prism, the spectrum is not produced directly. The operating principle, however, can be understood from a couple of examples.

Light from a monochromatic source will clearly form a simple interference pattern at the focus of the instrument. If the detector accepts only a narrow portion of that pattern, then its output will vary as the position of the moving mirror changes and different parts of the pattern impinge onto it. If the moving

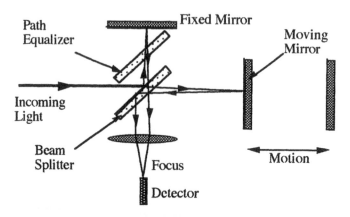

Figure 8.25 Optical components of a Fourier transform spectroscope.

mirror is scanned smoothly, then the detector output will be a simple sine wave (figure 8.26). Monochromatic light of a slightly different wavelength will give a similar output with a slightly different period. A single source in which both wavelengths are present will have these two outputs combined by straightforward addition, to give an output with a beat frequency (figure 8.27).

In frequency terms, a monochromatic source is a single delta function, and its Fourier transform (equation (8.17)) is a sine wave. A double delta function, corresponding to the bichromatic source, would have a Fourier transform with a beat frequency. We may see therefore that for more complex spectra, the output from the instrument will be related to the Fourier transform of the spectrum. Conversely, the spectrum may be found from the output through the inverse Fourier transform (equation (8.18)). Hence the name given to the instrument. The precise relationship between spectrum and output is obtained from the real part of the inverse transform:

$$I(\lambda) \propto \int_0^\infty I(\Delta P) \cos\left(\frac{2\pi \Delta P}{\lambda}\right) \mathrm{d}\Delta P \qquad (8.34)$$

where $I(\lambda)$ is the intensity in the spectrum at wavelength, λ, ΔP is the path difference between the two beams for a particular position of the moving mirror and $I(\Delta P)$ is the output (intensity) from the instrument for a particular value of ΔP.

Since there is no spectrum produced directly, there is no equivalent to the dispersion of a diffraction grating or prism. The spectral resolution of the Fourier transform spectroscope is theoretically infinite. In practice, however, the path difference cannot be extended to infinity as required by equation (8.34), and this, together with the measurements having to be made discretely instead of continuously, imposes a limit on the spectral resolution. The spectral resolution is thus given by

$$R = 2\Delta P_{\mathrm{max}}/\lambda \qquad (8.35)$$

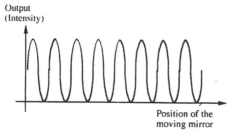

Figure 8.26   Output from a Fourier transform spectroscope for a monochromatic source.

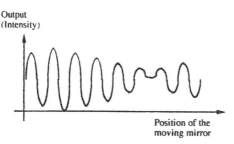

Figure 8.27   Output from a Fourier transform spectroscope for a bichromatic source.

where ΔP_{max} is the maximum path difference for the two beams that the instrument can achieve. For laboratory spectrometers, ΔP_{max} can be as much as two metres, giving a spectral resolution in the near infrared, where they are usually used, of over a million. For astronomical use, Fourier transform spectroscopes would normally be of smaller size and lower resolution than this, but some instruments used for solar spectroscopy do have resolutions of up to a hundred thousand.

The limit to the spectral resolution of a Fourier transform spectroscope means that there is no need to make continuous measurements of the output. This in any case is impossible, since even apparently continuous devices, like chart recorders, are in effect digital with sampling intervals given by their time constants. The sampling interval, and hence total number of measurements, for the Fourier transform spectroscope should be adequate to retain the spectral resolution, but not more than this or time and money will be being wasted. The sampling interval thus required is given by

$$\Delta x = \frac{(\lambda_1 + \lambda_2)^2}{16(\lambda_1 - \lambda_2)} \tag{8.36}$$

where Δx is the distance moved by the moving mirror between successive measurements of the output, and λ_1 and λ_2 are the short and long wavelength limits to the spectrum. Thus a Fourier transform spectroscope to obtain a spectrum from 1.0 to 1.2 μm at a spectral resolution of 10 000, would require a maximum path difference of 5.5 mm for the two light beams, and therefore a movement of the mirror by 2.75 mm. A total of 1800 measurements would be needed at 1.5 μm intervals for the moving mirror.

The Fourier transform spectroscope has two big advantages over most other types of spectroscope. These are known as the Multiplex (or Fellgett) and Throughput (or Jacquinot) advantages. The multiplex advantage occurs when spectroscopy in a particular wavelength region is limited by noise in the detector. As the name implies, this advantage of the Fourier transform spectroscope arises because the detector is viewing the whole spectrum all the time. The average intensity detected will be about half the integrated intensity of the whole spectrum. If a spectrum with N resolution elements is to be obtained, then the total time taken to obtain the spectrum if a grating or prism spectrum is scanned by the detector would be

$$T_{\text{scan}} = Nt \tag{8.37}$$

where t is the time taken to obtain a single measurement to the required signal to noise ratio. With a Fourier transform spectroscope, the same spectrum would require twice as many measurements (because the inverse transform gives both $I(\lambda)$ and $I(-\lambda)$), but the increased intensity being viewed by the detector would reduce the time needed to reach the required signal to noise ratio. The time for each measurement would thus be

$$\frac{t}{\sqrt{N/2}} \tag{8.38}$$

and the total time required to obtain the same spectrum using the Fourier transform spectroscope would be

$$T_{\text{FTS}} = \sqrt{8N}\, t. \tag{8.39}$$

The multiplex advantage is thus

$$T_{\text{scan}}/T_{\text{FTS}} = \sqrt{N/8}. \tag{8.40}$$

For the previous example of a spectrum from 1.0 to 1.2 μm, the Fourier transform spectroscope would obtain it in one fifteenth of the observing time required for a scanning instrument.

The multiplex advantage of the Fourier transform spectroscope disappears when the system is not limited by detector noise or if an array of detectors is used. The throughput advantage, however, remains. This advantage is the result of the larger entrance aperture allowed by the interferometer since no aperture is required to preserve spectral resolution; for a given resolution, the light grasp can be some two orders of magnitude larger than that of a grating or prism spectrometer.

There is a further advantage to the Fourier transform spectroscope which arises as a consequence of what at first sight seems to be one of its disadvantages.

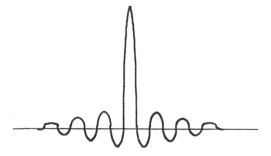

Figure 8.28   Natural instrumental profile of a Fourier transform spectroscope.

The Fourier transform spectroscope does not give a spectrum directly, but this has to be computed after the observations by using the inverse Fourier transform. Even with the fast Fourier transform algorithm, the delay between observation and spectrum can be significant. If observing from a remote site with inadequate computing facilities, the delay could be for days or more until the observer returns to base, thus eliminating the possibility of chasing up interesting details from the 'first look' at the data. The advantage that comes from the requirement of processing the output in order to obtain the spectrum is that it allows some image processing to be included in the data reduction. Of greatest significance in this respect is the ability to optimize the instrumental profile (or point-spread function) of the instrument. The natural instrumental profile of the Fourier transform spectroscope is of the form

$$\frac{\sin \Delta\lambda}{\Delta\lambda} \qquad (8.41)$$

where $\Delta\lambda$ is the distance from the central wavelength. This instrumental profile has fringes (figure 8.28), which can lead to spurious features appearing near very strong emission or absorption lines or edges. By weighting the data during the inverse transformation, the form of the instrumental profile can be changed, though usually at the cost of some slight reduction in the Rayleigh resolution. The commonest weighting function in use is triangular:

$$\omega(\Delta P) = 1 - \frac{\Delta P}{2\Delta x} \qquad (8.42)$$

which reduces the secondary maxima to those of a rectangular aperture, while the Rayleigh resolution deteriorates by a factor of two. Such a purposeful manipulation of the instrumental profile is part of the technique called Apodization.

8.8   FABRY-PÉROT ETALONS

The last form of disperser of current interest to the astronomer is the Fabry–Pérot Etalon. This is a classic disperser in the manner of the diffraction

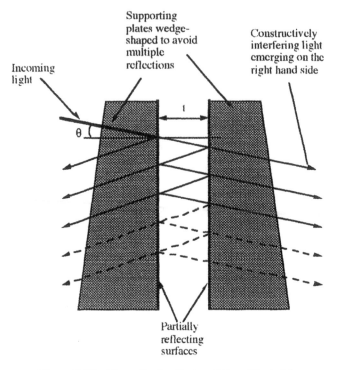

Supporting plates wedge-shaped to avoid multiple reflections

Constructively interfering light emerging on the right hand side

Incoming light

Partially reflecting surfaces

Figure 8.29 The optical paths in a Fabry–Pérot etalon.

grating or prism, producing a spectrum directly, and can be incorporated in their place into 'normal" spectroscope designs. It consists of two partially reflecting parallel surfaces (figure 8.29). Rays emerging from the right with integer number of wavelength path differences interfere constructively. Other wavelengths experience destructive interference, and so do not emerge on the right. Since the very first reflection is an internal one, and so has a 180° phase difference from the external reflections, the rays emerging on the left have a total destructive interference when there is constructive interference on the right hand side, and vice versa.

The path difference between successive rays is given by (figure 8.29)

$$\Delta P = 2t \cos \theta. \qquad (8.43)$$

The beam of light emerging on the right hand side will thus only contain those wavelengths for which ΔP is an integer number of wavelengths. The number of interfering beams, and hence the purity of the emerging wavelengths, depends upon the number of reflections. If the reflectivity of the partially reflecting surfaces is high then there will be many reflections and the purity of the emerging beams will be good (figure 8.30). Typically a reflectivity of 90% or so is used

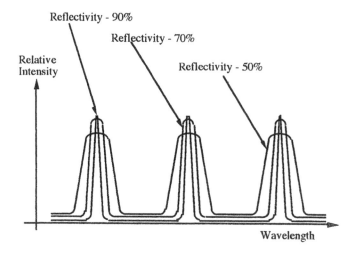

Figure 8.30 The output with respect to wavelength from a Fabry-Pérot etalon for a point source of white light.

in practical devices. The principal maxima shown in figure 8.30 will be at wavelengths given by

$$\lambda = \frac{2t \cos \theta}{m - 1}, \frac{2t \cos \theta}{m}, \frac{2t \cos \theta}{m + 1}, \cdots \qquad (8.44)$$

where m is an integer (the order of the spectrum). However, in the case of a point source, they will not be physically separated from each other, unless a cross disperser is used.

Using an entrance slit as the source, the angle of incidence onto the etalon will vary with position along the slit. For a monochromatic source, therefore, there will be a number of maxima accordingly as

$$\theta = \cos^{-1}\left(\frac{(m - 1)\lambda}{2t}\right), \cos^{-1}\left(\frac{m\lambda}{2t}\right), \cos^{-1}\left(\frac{(m + 1)\lambda}{2t}\right), \cdots \qquad (8.45)$$

corresponding to constructive interference at the one wavelength in different order spectra (figure 8.31), and these will be physically separated in the image.

A second monochromatic slit source with a different wavelength from the first would have a similar output to figure 8.31, with the physical positions of the maxima altered slightly. A bichromatic slit source would therefore have an output like that shown in figure 8.32. Finally a white light slit source would thus have an output from the etalon comprising a series of short spectra of differing orders superimposed onto each other. A cross disperser can then convert these into an array, and careful design of the system's parameters can lead to coverage of the complete spectrum, as for the echelle grating (above). The free spectral range of an etalon is given by equation (8.16).

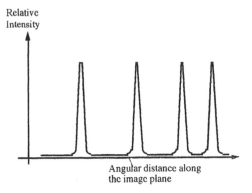

Figure 8.31 Output from a Fabry–Pérot etalon for a monochromatic slit source.

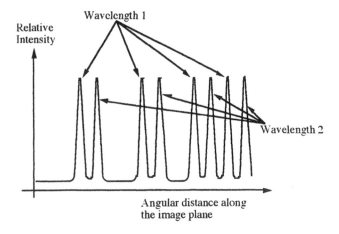

Figure 8.32 Output from a Fabry–Pérot etalon for a bichromatic slit source.

The dispersion of the etalon may be found by differentiating equation (8.44):

$$\frac{d\lambda}{d\theta} = \frac{2t\mu \sin\theta}{m} \tag{8.46}$$

where μ is the refractive index of the material between the plates of the etalon. From equation (8.44), and for the usual case of θ small and air between the plates, so that $\mu \approx 1$, we get

$$\frac{d\lambda}{d\theta} \approx \lambda\theta. \tag{8.47}$$

The equivalent of the Rayleigh spectral resolution for an etalon may be shown to be given by

$$R \approx \frac{2\pi t \sqrt{r}}{(1-r)\lambda} \tag{8.48}$$

where r is the fractional reflectivity of the surfaces of the etalon. In the laboratory, etalons may have spacings of several tens of centimetres and resolutions of several million. More modest instruments are in astronomical use, typically with separations of a few hundred microns and resolutions of a few times ten thousand or less.

9

Specialized Optical Spectroscopic Techniques for Astronomy

9.1 INTRODUCTION

Spectroscopy in the laboratory and in the observatory can sometimes be very closely related to each other, using the same instruments and techniques. More normally, however, spectroscopes designed for astronomy need to be optimized for that purpose (chapter 10), and different or additional techniques need to be used. We have already seen one major requirement, that of light gathering power or throughput. In terms of its utilization of the available light, the astronomical spectroscope must be as efficient as possible, because the sources are often very faint. Even those sources for which this would not seem to be a problem, such as the Sun, have been thoroughly studied in the past at high light levels, and current work uses very high spectral and/or angular resolutions so that light efficiency is again a fundamental requirement. This chapter looks at the other specialized requirements for spectroscopy as used in astronomy.

9.2 DETECTORS

Generally the same detectors can be used for astronomical spectroscopy as those used for direct astronomical imaging. The photographic emulsion is still to be found, especially in older instruments and for objective prism spectroscopy. However it is rapidly being replaced over the visible, near infrared and near ultraviolet by CCD detectors. These have a much higher quantum efficiency, even at the shorter wavelengths where fluorescent coatings have to be used. The image is fed directly into a computer where it may be easily processed (see later discussions). CCDs are, however, still relatively small, and so inappropriate for physically large spectra. Image Photon Counting Systems (IPCS) are also to be found and have the advantage of a continuous read-out of the image during exposure (see below). At wavelengths longer than about 1 μm, various array detectors based upon cooled semiconductor bolometers, some with charge

156

coupling read-out, are now in widespread use and have similar advantages to CCDs. In the ultraviolet, observations at wavelengths less than about 350 nm require the instrument to be lifted above the Earth's atmosphere. The main ultraviolet spectra to date have been obtained by the International Ultraviolet Explorer (IUE) spacecraft and the Hubble space telescope. The former uses an SEC vidicon TV camera as the detector within an echelle spectroscope operating down to 100 nm. The latter is discussed further in chapter 10.

9.3   GUIDING

One obvious difference between a laboratory and an astronomical source is that the latter is changing its position with time. Now all telescopes, except the cheapest ones aimed at the amateur market, are driven to counteract the effects of the Earth's rotation. So the changing position in the sky of the object under observation would not seem to matter for any spectroscope attached to or fed by such a telescope. Unfortunately the drives on most telescopes, including some of the largest and most modern, are inadequate to keep the stellar image centred on the entrance slit of the spectroscope, which may have a projected angular width on the sky of only 0.1″ or less. Even a drive which perfectly counteracted the Earth's rotation would not compensate for the changing refraction within the Earth's atmosphere as the zenith distance of the object altered, and so would fail to follow the object correctly throughout a long exposure. A system for correcting the imperfections in the telescope drive is therefore normally essential. This is generally known as guiding the telescope.

Guiding the telescope is normally essential for any purpose, not just for spectroscopy. For direct imaging, it is even more important, because any drift of the image will show up as a distortion (trail) of the image. In spectroscopy, drift will lead to longer exposures because the light from the object is not entering the spectroscope at all times, but without otherwise normally causing a deterioration of the spectral image. Guiding of a telescope is usually accomplished by viewing the object through a second telescope attached to the main one and adjusting the telescope's drive to keep the object centred on cross wires. This may be done visually, or there are numerous automatic systems fulfilling the same purpose. For very faint objects, off-set guiding may be necessary, when a brighter nearby object is used for guiding, and the guiding telescope off-set from the main telescope so that the object of interest remains centred in the main telescope. Alternative approaches include sampling a small portion of the field of view of the main telescope and guiding on an object in that region, or using a dichroic mirror to enable guiding to take place at one wavelength while the image is obtained at a different wavelength.

For some approaches to spectroscopy, such as multi-object slit spectroscopy, objective prism spectroscopy, or when the whole image of the object is accepted by the spectroscope, guiding has to be undertaken in the same manner as for

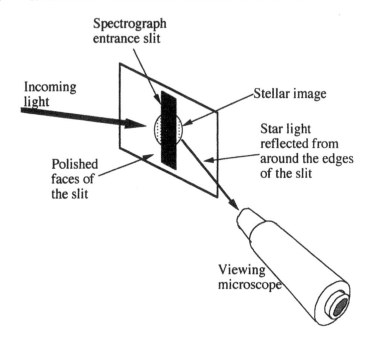

Figure 9.1   Use of a low power microscope to guide on the overspill for an image from the edges of the entrance slit to a spectroscope.

direct imaging. Much of the time, however, with single object slit spectroscopy, the image of the object, even if it is a star, will be larger than the entrance slit to the spectroscope. The overspill from the image may then be viewed via a low power microscope and used for guiding (figure 9.1).

9.4   WIDENING

The image of a star is not a point, nor even is it diffraction limited, because of spread caused by the Earth's atmosphere, arising from seeing, scintillation and twinkling. Nonetheless, a stellar image is normally only one or two seconds of arc across. If that image were to be guided to a fixed position on the slit, the resulting spectrum would be very narrow, and difficult to study. The spectrum is therefore normally deliberately widened to make it easier to analyse. This does, of course increase the required exposure in direct proportion to the increased width of the spectral image. Widening may be accomplished by setting the slit parallel to the diurnal motion of the object, and adjusting the telescope drive to be slightly too slow or too fast. The image will then drift slowly along the slit, and can periodically be reset to the other end of the slit using the drive's slow motion controls. More usually, however, the spectrum is widened using a separate device known as a rocker. This is a thick slab of glass with optically flat,

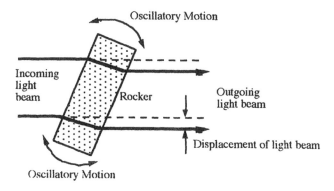

Figure 9.2   Use of a rocker to widen a spectrum.

parallel sides (figure 9.2). The slab is placed into the light beam and oscillated back and forth through a small angle. The varying displacement introduced into the beam by the slab then produces the required widened spectral image.

9.5   IMAGE DISSECTORS

The overspill from an image around the sides of the spectroscope entrance slit, though useful for guiding, represents a loss of light to the spectroscope and therefore results in an increased exposure. The light in the image can be more efficiently used through the use of an image dissector. This will also have the incidental benefit of widening the spectrum without requiring a separate mechanism. Image dissectors for modern spectrographs are made from a bundle of optical fibres. One end of the bundle is placed at the focus of the telescope and is matched in shape and size to the stellar image. The other end of the bundle feeds the spectroscope and is shaped to match its entrance slit (figure 9.3). Older devices used stacks of mirrors to the same end (figure 9.4).

9.6   DEKKERS

Most spectroscopes can change the available length of the slit by the use of an external stop. This will also usually have separate apertures to allow the comparison spectrum (see below) to be placed either side of the main spectrum without overlapping it. Such a stop is called a Dekker, and will usually have a range of lengths to allow for different objects and different observing conditions (figure 9.5).

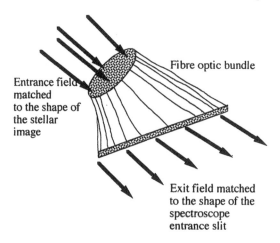

Figure 9.3   Image dissector formed from a fibre optic bundle.

9.7   LONG SLIT SPECTROSCOPY

The need for wideners and image dissectors is restricted to stellar spectroscopy. When an extended object is being observed, a much longer slit than normal can be used with advantage. The object needs to be guided to a fixed position on the slit to a high degree of accuracy. The final spectrum then gives, across its width, individual spectra for each portion of the object falling onto the slit (figure 8.20).

9.8   COMPARISON SPECTRA

9.8.1   Wavelength

From the earliest days of stellar spectroscopy it has been realized that a comparison spectrum would be needed to enable precise wavelength data to be obtained. Huggins made the first such attempt in 1868 when he compared the spectrum of Sirius with that of a hydrogen emission lamp. Few astronomical spectroscopes nowadays other than objective prisms and Fourier transform devices are without a means of producing such a spectrum. Indeed, one of the reasons for the use of a slit as the entrance aperture to a spectroscope is to facilitate the alignment of the comparison spectrum with the stellar (or other) spectrum.

The comparison spectrum is just an emission line spectrum, produced on or near the spectroscope from an emission lamp or an electric arc, the light from which is fed into the spectroscope. Usually the comparison spectrum is split and appears on either side of the spectrum of interest (figure 9.6). Measurement of the position of a spectral line (and hence its wavelength) on a photographic

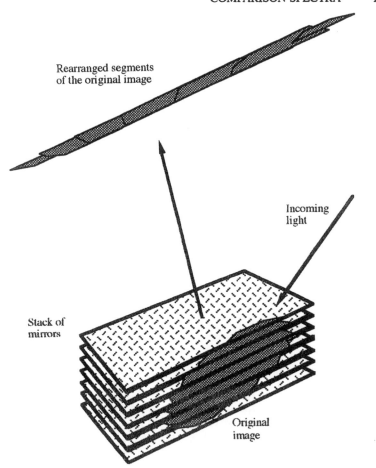

Rearranged segments
of the original image

Incoming
light

Stack of
mirrors

Original
image

Figure 9.4 Image dissector formed from a stack of mirrors (Bowen image dissector).

spectrum is made by comparison with the emission lines' positions using a high precision travelling microscope (figure 9.7). For a CCD or equivalent image, a similar process takes place upon the electronic image stored within a computer using appropriate data reduction software. Since the wavelengths of the emission lines are known, the observed wavelengths of lines in the spectrum of interest can be found by interpolation. A good deal of care needs to be taken if accurate wavelengths are to be obtained. For example with a photographic spectrum, the measurements need to be repeated several times and the spectrum rotated through 180° between each pass to reduce setting errors. During a single such measurement, the screw driving the travelling stage should rotate in only one sense, in order to eliminate errors due to backlash. With both photographic and electronic images, care has to be taken when lines are asymmetrical, the dispersion is nonlinear, and so on. With such care, however,

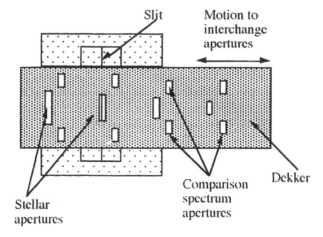

Figure 9.5 The use of a Dekker to limit the length of the entrance aperture of a spectroscope.

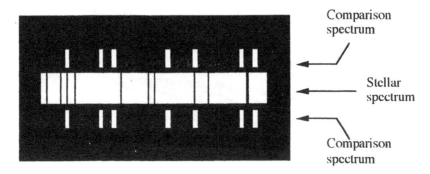

Figure 9.6 A stellar spectrum with a comparison spectrum.

observed wavelengths in the visible region may be determined typically to about one part in $30\,000/n$ (where n is the dispersion of the spectrum in nm mm^{-1}). This corresponds to being able to determine radial velocities through the Doppler shift to an accuracy of about $\pm 10n$ km s^{-1}.

9.8.2 Photometric

For detectors such as the CCD, which have a linear response, there is no problem in converting the detector output into relative spectral intensity. For nonlinear detectors, however, such as the photographic emulsion, a calibration curve is required to convert the signal into intensity. This curve is most usually obtained from a photometric comparison spectrum. A calibration curve, usually called the characteristic curve in the case of a photographic emulsion, may be obtained

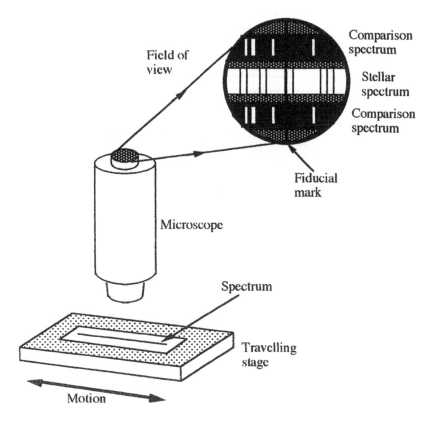

Figure 9.7 Use of the comparison spectrum to obtain the wavelengths of lines in the spectrum of interest.

in integrated light, but since the emulsion response may vary with wavelength, a calibration spectrum is more usual. This is a low dispersion spectrum with different sections across its width of different, and known, relative intensities (figure 9.8). From this a characteristic curve (figure 9.9) for the emulsion can be obtained, if necessary at a variety of wavelengths. Since the response of a photographic emulsion is highly variable with a number of factors, great care needs to be taken over the photometric calibration if accurate work is to be attempted. For example, the emulsion should be from the same batch as that used to obtain the spectrum, the exposure for the comparison should be of a similar length to that used for the main spectrum, the comparison and main spectra should be processed together etc. With such care errors in photographic spectrophotometry can be reduced to about 1%.

Although many of the difficulties of photography are avoided by the use of CCD and other array detectors with a linear response, there are still problems

Wavelength

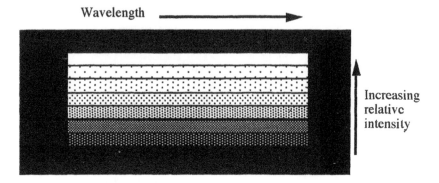

Increasing relative intensity

Figure 9.8 A photometric comparison spectrum—often emission lines may be added to facilitate identification of wavelengths along the spectra.

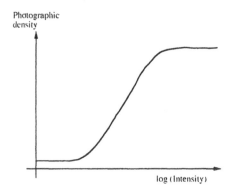

Figure 9.9 Schematic characteristic curve for a photographic emulsion.

to consider. The response of an individual pixel in an array may be linear, but the sensitivities of different pixels can vary. Furthermore there will be a dark signal (equivalent to the background fog on a photographic emulsion) and this may also vary from pixel to pixel. These effects have to be removed by taking exposures with similar lengths to that of the main exposure, firstly while observing a uniformly illuminated surface (flat fielding) to find the varying responses of the pixels, and secondly without any light incident onto the detector (dark frame subtraction) to reduce the effect of the dark signal. Since charge transfer in a CCD is not 100% efficient, a very bright part of the image will give spuriously high readings in adjacent pixels during the read-out process. This problem is of more importance in direct images where one or two stars may be very much brighter than the rest, but it is not negligible in spectrophotometry, especially when there are strong emission lines in the spectrum. There may also be 'bad" pixels, which have a poor response, very high dark signal or poor charge transfer, whose effects cannot be completely corrected by the flat fielding or dark frame subtraction. Then there are the effects of cosmic rays upon any of the three exposures. A cosmic ray passing through the detector during an exposure will ionize a large number of the atoms, and the resulting·electrons will be gathered up in the same way as those produced by the photons contributing to the image. One or two pixels around the point of entry of the cosmic ray will thus have an anomalously high signal. Such pixels can usually be recognized by careful inspection of the image, and as a partial remedy, the spurious signal replaced by the average of the eight surrounding pixels. Unfortunately, at the time of writing, correction of cosmic ray spikes is a highly tedious process which usually still has to be done by hand.

9.9   FLEXURE

The components of an astronomical spectroscope must be kept in their correct relative positions if the optical performance of the instrument is not to be degraded. With the exception of instruments placed at the Coudé focus of a telescope (or with an alt-az type mounting, at one of the the Nasmyth foci), the orientation of the spectroscope will vary as the telescope is pointed to different parts of the sky. The instrument must thus be designed so that the flexure arising from the changing gravitational loadings on it does not alter the positions and orientations of its components beyond the point at which the performance of the instrument would start to deteriorate. Early spectroscopes followed closely the design of the basic spectroscope (figure 8.14), and were attached in place of the eyepiece. They were therefore poorly supported and suffered badly from flexure. To reduce the problem the practice of reflecting the light beam from the telescope through 90° before incidence onto the spectroscope was soon adopted. The plane of the spectroscope is then the same as that of the primary mirror. A substantial mount for the spectroscope can thus be attached at several points to the equally substantial mounting for the primary mirror, reducing flexure to manageable proportions. Most modern spectroscope designs follow this practice, but care is still needed in their design to minimize the remaining flexure. The use of a folded design for the spectroscope, such as the Wadsworth (figure 8.18), can often be beneficial.

9.10   TEMPERATURE

Except for spectroscopes at the Coudé focus of a telescope, which may be kept in a temperature controlled room, most instruments are attached to the telescope, and so must be at or close to the ambient temperature. This is usually low and will vary throughout the night. The relative positions of the spectroscope's components can therefore alter through the thermal expansion or contraction of their supports. The use of suitable materials in the construction of the instrument, such as Invar can usually eliminate this difficulty. If any problems remain, then the spectroscope itself can be heated to keep it at a constant temperature a few degrees above the ambient temperature. However, this practice can reduce the quality of the images through the production of convection currents near the telescope.

More seriously, the dispersing element may be affected by temperature changes. If it is a diffraction grating, then the use of zerodur, ULE or other low expansion material for its construction will reduce the problem almost to zero. The dispersion of a prism, however, is quite temperature dependent. If the spectral resolution is not to be degraded, then the temperature of the prism must be kept constant to about 0.1 K throughout the exposure. This can only be accomplished by the use of a precision thermostat and heating jacket.

9.11 EXPOSURES

The exposure required for a spectral image clearly depends upon the throughput, the widening and the detector sensitivity. The effects of these factors can be determined, at least approximately, in advance. However, spectroscopy, unlike much work in astronomy, can continue when the sky transparency is poor, or even through thin cloud. Furthermore, an exposure interrupted by cloud can be continued once the clouds have gone. It is even possible, given a stable enough spectroscope, to continue an exposure the following night; Edwin Hubble in his pioneering work on the red shift of the galaxies took exposures which extended over several complete nights. In such circumstances, it is clearly difficult to judge the correct exposure needed, and this is complicated further by the required exposure varying with the purpose for which the spectrum is needed. If strong emission lines are to be studied, then a much shorter exposure can be given than if faint absorption lines are of interest. Spectra from which radial velocities are to be measured can be of a lower signal to noise ratio than those from which line profiles are desired, and so on. For some purposes, Bowen's formula may be of use as a guide to the limiting magnitude for a spectroscope in which the star's image is larger than the slit width:

$$m = 12 + 2.5 \log_{10} \left(\frac{s D_1 T_D g q t \, (d\lambda/d\theta)}{f_1 f_2 H \alpha} \right) \tag{9.1}$$

where m is the faintest B magnitude which will give a usable spectrum in an exposure of t seconds, s is the projected slit width, D_1 is the diameter of the light beam passing through the dispersing element, T_D is the diameter of the telescope objective's, g is the optical efficiency of the system (i.e. the ratio of the usable light at the focus of the spectroscope to that incident upon the telescope)—typical values would be in the range 0.1 to 0.5, q is the quantum efficiency of the detector, $d\lambda/d\theta$ is the dispersion of the spectrum, f_1 and f_2 are the focal lengths of the collimator and imaging element respectively, H is the height of the spectrum and α is the angular size of the stellar image at the slit. With some detectors, such as the IPCS, there can be a continuous display of the spectrum throughout the exposure, and it can simply be continued until the desired signal to noise ratio is achieved. With CCDs, it is easy to add separate exposures together, though this does increase the read-out noise. If an image is read out, and is found to be underexposed, then a second or third exposure can be added, again until the desired signal to noise ratio is achieved. For photographic spectroscopes it is less easy, though not impossible, to add images together.

Some instruments have an integrating exposure meter. This samples a small part of the spectrum and integrates the signal. The exposure can then be stopped when an adequate exposure has been given. Much of the time, however, no such devices are available, and the observer has to act as his or her own exposure meter, and estimate appropriate allowances for sky transparency, cloud etc.

10

Examples of Optical Spectroscopes

10.1 INTRODUCTION

In this chapter some examples of actual astronomical spectroscopes are examined. The danger of reviewing current instruments in a book, of course, is that they may be out-of-date by the time the book appears, or soon afterwards. Nonetheless it is still worth looking at current instruments for several reasons. Firstly, the development of spectroscopes has been occurring for well over a century, and current examples are therefore close to the limits of what can be achieved until some very radical new development occurs. Secondly, and partly because of the first reason, individual instruments tend to remain in use for long periods with only minor modifications. The author for example has used spectroscopes constructed in the early 1950s that are still to be found on some major telescopes. Finally, and again because of the first reason, even if new instruments do come into use, they are unlikely to be fundamentally different from existing designs.

No examples of prism-based slit spectroscopes have been included, because, whilst some such instruments may remain in use, it is now highly unlikely that any new spectroscope for a major telescope would be designed purely using a prism as the dispersing element. Designs utilizing objective prisms and grisms (combined gratings and prisms) are covered, however.

10.2 A SMALL BASIC SPECTROSCOPE

A simple grating-based slit spectroscope, designed for use on a small telescope and primarily intended for teaching purposes, is shown in figure 10.1. Its optical layout (figure 10.2) clearly shows all the elements of the basic spectroscope (figure 8.14), and its parameters are listed in table 10.1. Such an instrument on a moderate telescope is capable of providing reasonable spectra of the brighter stars, sufficient for spectral classification, monitoring spectrum variables, for element identification etc.

168

Figure 10.1 A simple basic spectroscope attached to a 0.4 m telescope (Optomechanics™ model 10C).

10.3 A CONVENTIONAL CASSEGRAIN SPECTROSCOPE

The workhorse of spectroscopy is a conventional grating-based slit spectroscope attached to the Cassegrain focus of a large telescope. Its size is limited by the space available at the back of the main mirror mounting of the telescope, and its weight is limited by the strength of the telescope's tube and mounting. There are numerous examples of this type of instrument, only one of which, ISIS, is discussed here.

ISIS (Intermediate dispersion Spectroscopic and Imaging System) is one of the most frequently used spectroscopes for the 4.2 m William Herschel telescope on La Palma. It is fixed at the Cassegrain focus, and is reasonably representative of this type of instrument except that it is a dual system. It uses a dichroic mirror to split the incoming light into a red beam and a blue beam, each of which is then separately directed into a spectroscope optimized for those wavelengths (figure 10.3). The slit is common to both beams and can be varied from 50 μm to 5 mm in width (0.2″ to 23″ on the sky). It is followed by a dichroic mirror which separates the incoming white light into the blue beam (reflected) and the red beam (transmitted). Several mirrors are available to give the separation point at various wavelengths in the range 450 to 550 nm. The parameters of the system are listed in table 10.2.

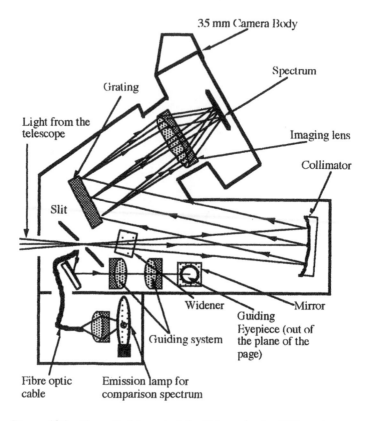

Figure 10.2 The optical layout of the Optomechanics 10C spectroscope.

Table 10.1 Parameters of the Optomechanics model 10C spectroscope.

Slit width	50 or 100 μm
Collimator	Focal length 225 mm, $f9$
Gratings	Reflection grating, 600 lines per mm, blazed for 500 nm in the first order, dispersion 12 nm mm^{-1}. Spectral resolution 500 or 1 000
	Reflection grating, 1200 lines per mm, blazed for 500 nm in the first order dispersion 6 nm mm^{-1}. Spectral resolution 1 000 or 2 000
Imaging element	Focal length 135 mm, $f2.8$
Comparison spectra	Fe/Ar or Hg/ Ne

10.4 TRANSMISSION GRATING SPECTROSCOPES

The use of a transmission grating is much less common than the use of a reflection grating in astronomical spectroscopes, but some examples do exist.

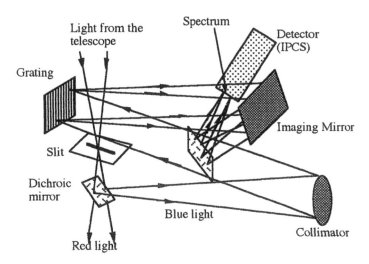

Figure 10.3 Schematic optical layout of the blue arm of ISIS; the red arm is similar and placed underneath the blue arm.

Table 10.2 Parameters of ISIS.

Slit width	50 μm to 5 mm
Collimators	Each of focal length 1.65 m, producing beams 150 mm in diameter
Gratings	Various reflection gratings, each 154× 206 mm in size, with from 150 to 2400 lines per mm, giving dispersions from 0.8 to 12 nm mm^{-1}. The blaze wavelengths range from 360 to 720 nm, and the spectral resolution from 800 to 7000
Imaging elements	Each of focal length 500 mm, f1.0
Detectors	For the blue beam, an IPCS with a CCD as the detector; for the red beam, a CCD
Comparison spectra	Cu/Ar, Cu/Ne, Th/Ar, or a continuum
Guiding	An auto-guider, a TV system for viewing the slit, or a blind offset from a brighter guide star
Optional elements	A multislit or fibre optic feed, a cross disperser, a polarizer, neutral density and colour filters (the latter to eliminate overlap between orders)

Thus, ISIS (above) has a faint object spectroscope (FOS) attached to it as a third arm, which relies on a transmission grating for its main dispersion. In the FOS, the grating is cemented to a direct vision prism which acts as a cross disperser (such a combination is known as a Grism, see below), in order to separate the first and second order spectra.

Here we look at another example of a transmission grating-based spectroscope, the Low Dispersion Survey Spectroscope (LDSS). This instrument

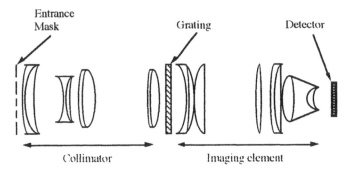

Figure 10.4   Optical layout of LDSS (schematic).

Table 10.3   Parameters of LDSS

Slits	Up to 100 slits 50 μm wide (plus long slit, no slit and direct options)
Gratings	Two, giving dispersions of 16.5 and 87 nm mm^{-1} with spectral resolutions of about 400 and 70, respectively
Detectors	IPCS or CCD
Spectral range	370 to 750 nm
Throughput	About 60%

is designed to obtain the spectra of up to 100 objects simultaneously at low dispersions, and down to about magnitude 23 when attached to the 3.9 m Anglo–Australian telescope (AAT). For greatest efficiency, the spectroscope uses lenses throughout in order to avoid the obstructions caused by the secondary mirrors in reflecting systems. All the lenses have multilayer coatings to reduce reflection losses. The entrance aperture can take several forms. In the principal mode of operation the entrance aperture is a mask with up to 100 precisely positioned slits to match the positions of objects in the field of view. Each mask is unique and has to be made specifically for each field of view observed with the instrument. Because of the width of the entrance field, the optics of this spectroscope need to have their aberrations corrected over a much wider field of view than those for normal spectroscopes, and this requirement is a second reason for the choice of transmission optics. The optical layout of the spectroscope is shown in figure 10.4, and its parameters listed in table 10.3.

10.5   COUDÉ SPECTROSCOPES

The weight and size constraints imposed on a Cassegrain spectroscope are considerably relaxed for a spectroscope placed at the Coudé or Nasmyth focus of a telescope. The problems of flexure are also eliminated or much reduced, because the spectroscope does not change its attitude as the telescope moves.

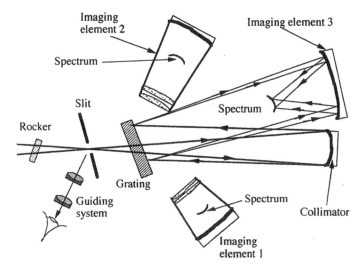

Figure 10.5 The Coudé spectroscope on the 1.52 m telescope of the European Southern Observatory (ESO) at La Silla, showing the light paths to the third imaging element.

The image from the telescope can often be brought into an area separate from the main observatory through an optical window. This area can then be temperature controlled to suit either the operators or the spectroscope (or both), reducing the problems due to temperature variations and improving observing conditions. Coudé and Nasmyth spectroscopes can thus be much larger, have higher dispersion and/or resolution, and be less robustly constructed than Cassegrain spectroscopes. The disadvantages of a Coudé or Nasmyth focus are that the image rotates as the telescope tracks the object across the sky, and that the focal ratio is usually very long.

As an example of this type of instrument, we look at the Coudé spectroscope on the 1.52 m telescope of the European Southern Observatory (ESO) at La Silla. This instrument uses reflection gratings in a conventional design. Three separate imaging elements are incorporated into the one mounting and can be selected by changing the angle of the grating. The optical layout of the spectroscope is shown in figure 10.5, and its parameters listed in table 10.4. Note the much better dispersion and spectral resolution possible with this instrument compared with ISIS.

10.6 A GRISM-BASED SPECTROSCOPE

The combination of a transmission grating and a prism, known through an unpleasant contraction as a Grism, can sometimes offer advantages over a pure grating-based system. This is particularly the case when a high efficiency instrument is required for the study of the spectra of very faint objects. The Faint

Table 10.4   Parameters of the Coudé spectroscope on the ESO 1.52 m telescope.

Slit width	50 to 300 μm (0.2″ to 1.3″ on the sky)
Collimators	Focal length 6 m, f30
Gratings	Various reflection gratings, each about 200 $\times$ 300 μm in size, with from 770 to 1500 lines per mm, giving dispersions from 0.26 to 3.1 nm mm^{-1}. The blaze wavelengths range from 410 to 750 nm, and the spectral resolution from about 10 000 to 100 000
Imaging elements	Three, of focal lengths 0.41, 0.67 and 0.25 m, and focal ratios 1.0, 1.6 and 2.2, respectively
Comparison spectra	Fe, Ne , Ar, or a continuum
Options	A field rotator to compensate for the rotation of the field of view due to the tracking motion of the telescope, and an exposure meter which uses the light blocked by the plate holders

Object Spectroscope (FOS) of the William Herschel telescope is an example of such a spectroscope. It is designed for low dispersion spectroscopy over a wide spectral range, and achieves this by using both the first and second order spectra produced by the grating and separating them by a direct vision prism acting as a cross disperser (figure 10.6). The two spectra overlap to give continuous coverage from about 350 nm to 970 nm. The optical layout of the instrument is shown in figure 10.7 and its parameters in table 10.5. In order to improve efficiency, the number of surfaces is kept to a minimum, so no collimator is used, and the imaging element is a Schmidt camera. The instrument is attached to the ISIS spectroscope (see above) at the Cassegrain focus of the WHT, and shares some of its components. It can be used simultaneously with the blue arm of ISIS.

10.7   MULTI-OBJECT SPECTROSCOPES

The LDSS (see above) can obtain the spectra of up to 100 objects simultaneously and is one example of this class of instrument. It requires, however, a mask containing slits positioned to match the objects to be studied and this has to be made up for every observation. A more flexible system, which also does not require the large field required for the optics of LDSS, is to use fibre optic cables to bring the light from the stars in the field of view to the slit of the spectroscope. The fibre optic cables can be moved around the image plane to intercept the light from the required objects easily compared with the labour required for machining a precision mask. A number of such instruments have been brought into use in the last decade, of which the ESO's MEFOS (Meudon–ESO Fibre Optics Spectroscope) is taken as an example.

MEFOS is positioned at the prime focus of the ESO's 3.6 m telescope, and feeds a spectroscope at the Cassegrain focus. The fibre optic cables are 21 m

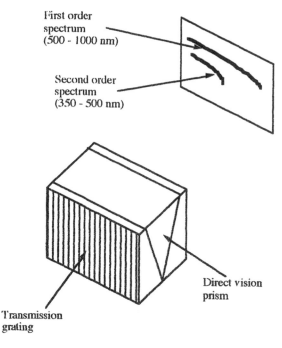

Figure 10.6   The Grism of the William Herschel Faint Object Spectroscope.

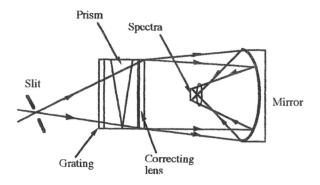

Figure 10.7   The optical layout of the William Herschel Faint Object Spectroscope.

long and the total light losses introduced by MEFOS are estimated at about 20%. It can observe up to 30 objects simultaneously over a 1° field, and the fibre optics are moved to the calculated positions of those objects by computer-controlled positioning arms (figure 10.8). Three fibres are carried by each arm. One intercepts the light from the object to be observed, the second the light from the nearby sky to provide background subtraction; these two fibres each cover

Table 10.5   Parameters of FOS.

Slit	50 μm to 5 mm or 0.2″ to 23″ on the sky (shared with ISIS to which the FOS is attached, see above). Two slits are used in order to obtain a sky background simultaneously with the stellar spectrum
Collimator	None. The system accepts the f11 beam from the focus of the WHT directly
Grating	A plane transmission grating, blazed at 730 nm, with 150 lines per mm cemented to a direct vision prism as a cross disperser. The first order spectrum covers from 460 to 970 nm at a dispersion of 40 nm mm^{-1}. The second order spectrum covers from 350 to 490 nm at a dispersion of 20 nm mm^{-1}. The spectral resolution is about 500 in the first order and 1000 in the second order
Imaging element	An $f1.4$ Schmidt camera accepting an $f11$ incoming beam
Detectors	A CCD detector
Throughput	17% including telescopic and atmospheric losses

2.6″ on the sky. The third (actually a fibre optic bundle covering 36″ × 36″ on the sky) provides a separate direct image for positioning purposes. The $f3$ beam from the telescope feeds with high efficiency directly into the fibres. No front lenses or prisms are required as can be the case with some other comparable systems. To start an observation, the positioning arms first move the imaging bundles to the calculated positions of the desired objects (figure 10.9). The actual alignments of the imaging bundles with respect to their targets are then found by taking a short direct exposure. The positions of the arms are corrected to centre the targets and are then offset to align the images precisely onto the spectroscopic fibres. The light from the objects is then fed into a spectroscope for the actual exposure. Guiding is accomplished by setting one of the imaging fibres onto a convenient bright star in the field. The system can be used with most spectroscope designs provided the collimator is matched to the $f3$ beam emerging from the fibre optics.

10.8   AN ECHELLE GRATING SPECTROSCOPE

An echelle grating can give high dispersions and spectral resolutions, but in order to achieve these it usually has to be used at high spectral orders and so must be allied to a cross disperser. The resulting spectrum is then a two-dimensional array of short segments of spectra. A major consideration in the design of an echelle spectroscope is therefore to ensure overlap at the ends of successive segments so that the whole spectrum may be reconstituted. As an example of an echelle grating based spectroscope, we look at those on board the International Ultraviolet Explorer (IUE) spacecraft.

The IUE spacecraft is designed specifically to obtain spectra in the ultraviolet.

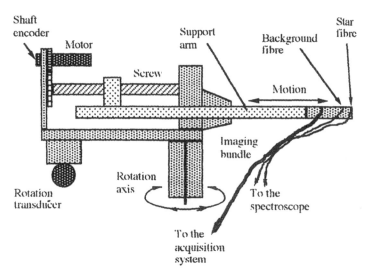

Figure 10.8   Schematic side view of a position arm from MEFOS.

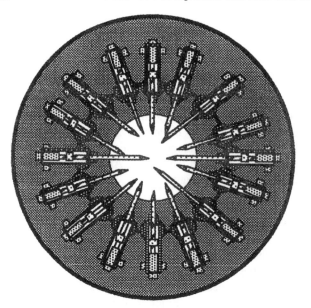

Figure 10.9   Schematic view of MEFOS (only 16 of the 30 positioning arms actually shown).

It carries a 0.45 m, $f15$ Ritchey–Chretien telescope which has two spectroscopes at its 'Cassegrain' focus. These spectroscopes cover the wavelength ranges

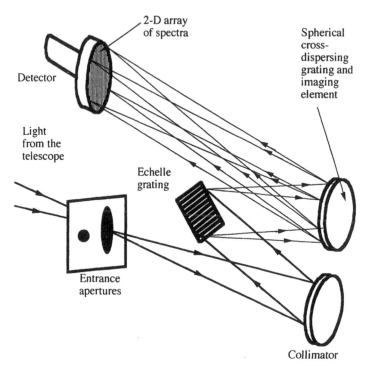

Figure 10.10 The optical layout of one of the spectroscopes on board the IUE spacecraft.

115–195 nm and 190–320 nm. Both spectroscopes use an echelle grating as the primary disperser, and then a low dispersion spherical grating which acts both as the cross disperser and as the imaging element (figure 10.10). The parameters of the spectroscopes are listed in table 10.6. An example of a high dispersion spectrum from IUE is shown in figure 8.10.

10.9   INFRARED SPECTROSCOPES

Full coverage of the infrared region requires that the spectroscope be lifted above most or all of the Earth's atmosphere, and balloon, rocket and spacecraft-borne spectroscopes have been so used. The atmosphere does, however, have some gaps in its near-infrared absorption spectrum, allowing radiation to reach the surface. Thus ground-based instruments, especially those working from dry sites at high altitude, can still be of use. The main difference between such an infrared spectroscope and those operating in the visible, apart from the obvious requirement for the optical components to transmit or reflect at the desired wavelengths, is the need to reduce background noise. Since the components of the telescope and spectroscope themselves can be emitting radiation in the

Table 10.6   Parameters of the IUE spectroscopes.

Entrance aperture	100 μm circular or 330 $\times$ 660 μm elliptical (3″ or 10″ $\times$ 20″ on the sky)
Collimators	Each off-axis paraboloids of focal length 1.89 m
Gratings	Short wavelength echelle grating; 102 lines per mm covering the range 115–195 nm over spectral orders 66 to 125 at a dispersion ranging from 0.086 to 0.137 nm mm^{-1}, and a spectral resolution of 12 000. The spherical grating used as a cross disperser has 313 lines per mm, and a focal length of 0.69 m.
	Long wavelength echelle grating; 63 lines per mm covering the range 190–320 nm over spectral orders 72 to 125 at a dispersion ranging from 0.125 to 0.2 nm mm^{-1}, and a spectral resolution of 13 000. The spherical grating used as a cross disperser has 200 lines per mm, and a focal length of 0.69 m. In both spectroscopes a plane mirror can be inserted in front of the echelle grating, leaving the cross disperser as the only dispersing element and giving a low dispersion mode to the system. The dispersions and spectral resolutions are then respectively 6 nm mm^{-1} and 250 for the short wavelength spectroscope, and 6 nm mm^{-1} and 300 for the long wavelength spectroscope
Detectors	An UV–optical converter followed by an SEC–vidicon television camera
Throughput	About 2% for the short wavelength spectroscope and from 0.45% to 4% for the long wavelength spectroscope (including the losses in the telescope)

infrared, this requirement usually translates into cooling some or all of the instrument. The ESO's infrared spectrometer, IRSPEC, is chosen as an example of this class of instrument.

This instrument was designed for use initially at the Cassegrain focus of the ESO's 3.6 m telescope, and later at the Nasmyth focus of the 3.5 m New Technology Telescope (NTT). The optical layout of the spectroscope is of relatively conventional design (figure 10.11). It is, however, encased in a vacuum chamber and cooled with continuously circulating liquid nitrogen to about 80 K, with the detector further cooled to 50 K by a thermal finger from solid nitrogen in a cryostat inside the main vacuum chamber. The optical components are mounted on an uncooled support from which they are thermally isolated, so that they may be aligned at room temperature and then cooled without going out of alignment. The parameters of the spectroscope are listed in table 10.7.

10.10   SPACECRAFT-BORNE SPECTROSCOPES

Any spectroscope operating at wavelengths shorter than about 350 nm, in the longer infrared or microwave regions, has to be lifted above the Earth's

Table 10.7 Parameters of IRSPEC

Aperture	800 μm (6″ on the sky)
Collimator	A 100 mm, f7.4 off-axis paraboloid
Gratings	Various reflection gratings, each 120 × 150 mm in size, covering the wavelength range 1–5 μm at spectral resolutions from 1 000 to 2 500
Imaging element	An off-axis, f2 Pfund-type camera
Detectors	A cooled infrared array
Comparison spectra	Ne or Kr
Guiding	Via a TV system for viewing the slit

atmosphere to make its observations. In other parts of the spectrum, being above the atmosphere can be a considerable benefit through the absence of absorption bands. At any wavelength, the observations will gain from the elimination of atmospheric image degradation (seeing), the reduction in background light, the absence of clouds and the reduction in gravitational loading. Thus many spectroscopes have been mounted onto spacecraft, rockets, balloons or very high flying aircraft to gain some or all of these advantages. One such spectroscope, that on board the IUE spacecraft, has been discussed in detail already. Here we look at another example, the Goddard High Resolution Spectroscope (GHRS) on the Hubble Space Telescope (HST).

The GHRS is designed to produce high quality spectra over the wavelength range 110–320 nm. The limits are due to the magnesium fluoride coatings on the telescope's optics and to the detector limits, not to the design of the spectroscope. Its optical layout (figure 10.12) is conventional but with a large number of options. Thus it may be used with conventional reflection gratings and a separate imaging mirror, or with an echelle grating and a concave cross-dispersing grating which also acts as the imaging element. Three of the conventional gratings are holographic and one ruled. The instrument also has two channels covering the ranges 110–180 nm and 115–320 nm. Exposures can be as short as 50 ms, and repeated rapidly for observation of changing sources. The longest single exposure is limited to about five minutes by motions within the instrument arising from the Earth's magnetic field. Longer exposures must be made by adding together several shorter exposures. Most objects are visible for about 45 minutes between 'rising' and 'setting' due to the spacecraft's orbital motion. Thus the longest exposures must extend over several orbits. The parameters of the instrument are listed in table 10.8.

10.11 FABRY–PÉROT SPECTROSCOPES

Fabry–Pérot etalons are most likely to be found in an observatory in use as narrow band filters; their use in spectroscopes is more usual in laboratory-

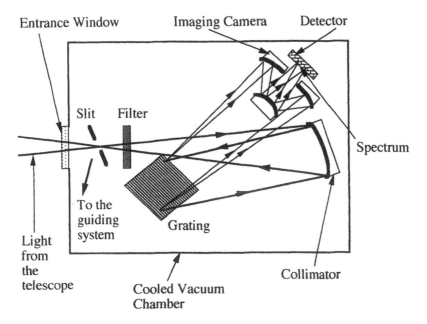

Figure 10.11 Schematic optical layout of IRSPEC.

Table 10.8 Parameters of GHRS.

Apertures	72 and 560 μm (0.22″ and 1.74″ on the sky)
Collimator	A 80 mm f23 off-axis paraboloid
Gratings	Four conventional reflection gratings, used in the first order, with from 600 to 6 000 lines per mm, covering the wavelengths from 110 to 320 nm at spectral resolutions from 2000 to 25 000. An echelle grating with 316 lines per mm, used at orders from 17 to 53, over the same wavelength range, at a spectral resolution of 80 000. Two cross-dispersing concave gratings with 195 and 86 lines per mm, and focal lengths of 1.46 and 1.34 m
Imaging elements	Two alternative concave mirrors, 84 mm, f17 and 86 mm, f15.7, respectively
Detectors	Two Digicon detectors, one with a caesium iodide photocathode and lithium fluoride window, detecting only wavelengths shorter than 180 nm, and the other with a caesium telluride photocathode and magnesium fluoride window for longer wave detection

based instruments. Nonetheless, there are a few astronomical examples, and we consider here TAURUS, which additionally serves as an illustration of the desperation with which acronyms may be sought since its name is derived from the 'Taylor–Atherton Uariable-resolution Radial Uelocity System'!

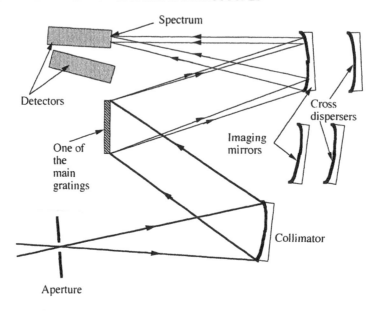

Figure 10.12 Schematic optical layout of the GHRS, showing just one of the many light paths that it is possible to select.

The second generation TAURUS-2 instrument was designed for the 3.9 m Anglo–Australian Telescope (AAT). It is intended primarily for mapping the velocities of emission-line objects. The instrument has the etalon placed in a collimated beam from the telescope. A filter isolates one of the transmitted orders, and the beam is re-imaged. The final image, detected by an IPCS or CCD detector, is thus pseudo-monochromatic. That is, at any point in the image the radiation covers only a very narrow waveband, but because of the changing angle of incidence onto the etalon, the centre of that waveband changes over the image. In operation, the spacing of the etalon is varied by piezoelectric spacers to scan its transmitted wavelength over the free spectral range of the order being used. Up to a hundred images may be observed. An image-cube is then reconstructed from the data with the image in two of the dimensions and the spectral dispersion orthogonal to this. The wavelengths across a single image have to be appropriately binned to allow for their variation over the field of view. The order to be observed is selected to be centred on one of the emission lines in the object. The velocities within the object observed can then be obtained from the change in the wavelength of the chosen emission line. The optical layout is shown in figure 10.13 and its parameters in table 10.9.

10.12 FOURIER TRANSFORM SPECTROSCOPES

These are even more rarely found in astronomy than Fabry–Pérot instruments, but they do occur in a few very specialist applications. Two of these few

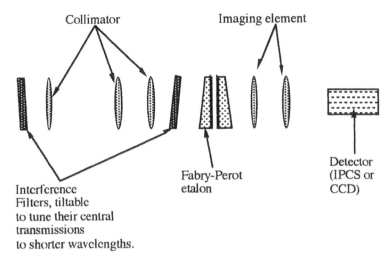

Figure 10.13   Schematic optical layout of TAURUS.

Table 10.9   Parameters of TAURUS.

Field of view	9 minutes of arc
Collimator	60 mm $f11$ lens system
Etalons	Various, with gaps ranging from 15 to 500 μm, velocity resolutions ranging from 7 to 150 km s^{-1}, spectral resolutions ranging from 500 to 60 000, spectral orders typically a few hundred, and covering the range 360 to 945 nm
Imaging elements	Two, with 61 mm, $f2.1$ and 61 mm, $f4$ lenses
Throughput	Up to 30%

examples are the Fourier Transform Infrared Spectroscopes (FTIS) on board the two Voyager spacecraft. The instruments were used, amongst other purposes, to identify compounds in Jupiter's, Saturn's and Titan's atmospheres, and to determine the composition of Martian clouds. The instruments had to be kept at 200 ± 0.5 K, and were therefore provided with thick thermal shielding and three sets of heaters, as well as radiators for cooling. They also had to be radiation-hardened in order to survive passage through Jupiter's radiation belts. The optical layout is shown in figure 10.14. In addition to the main Fourier transform spectrometer (FTS) shown in figure 10.14, the instrument had a separate reference FTS observing the 585 nm line from an on-board neon source, and the spherical dichroic mirror allowed visible light through to a separate radiometer.

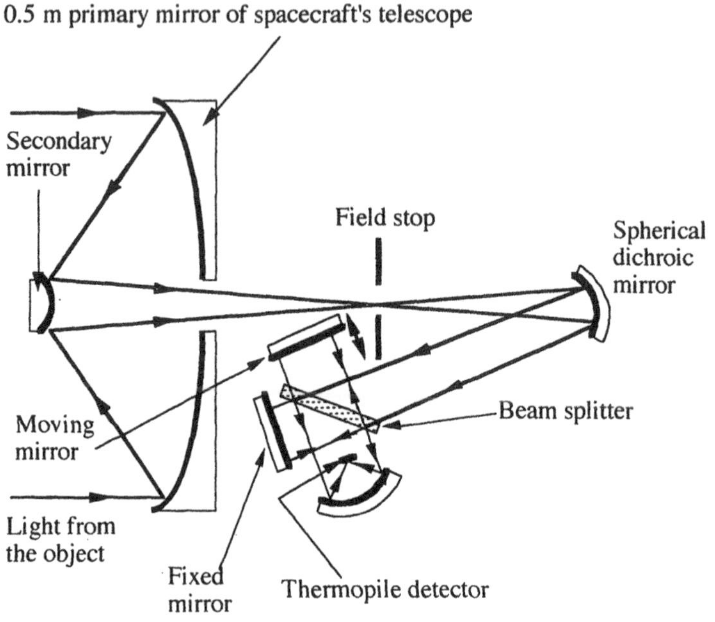

Figure 10.14 Optical layout of the FTIS on board the Voyager spacecraft.

10.13 OBJECTIVE PRISM SPECTROSCOPES

The simplest spectroscope of all is the objective prism. This (chapter 8) is just a large, low angle prism covering the whole aperture of the telescope, so that every object on the image is represented by its spectrum. Objective gratings can also be found, but these are for assisting photometric and astrometric measurements, not for spectroscopic purposes. Some variants of objective prisms replace the single large prism with a matrix of co-aligned smaller prisms in order to reduce costs, but this is at the expense of throughput because of the obstruction to the aperture caused by the prism mountings. Other designs use two opposed prisms of differing glasses to provide a direct vision prism (figure 8.23). While this avoids the offset otherwise required to set the telescope onto an object, it reduces the dispersion of the spectrum. The spectral resolution of an objective prism can be theoretically very high because of the large size of the prism (equation 8.33). In practice, however, it will normally be limited by the spatial resolution of the detector because the spectra are physically small due to the low dispersion of the narrow angle prism. On conventional telescopes, objective prisms are being replaced by multi-object instruments (see above) which have higher dispersions and spectral resolutions, can easily provide a wavelength calibration and can use the efficient CCD or IPCS as the detector. Objective prisms and photographic emulsions are still likely to be used for survey purposes for some time to come

Table 10.10   Parameters of the Objective Prisms for the UK Schmidt Telescope.

Apex angle	Two full aperture prisms with apex angles of 0.73° and 2.25°
Dispersion	The prisms can be used individually or in combination. In the latter case they can be aligned so that their dispersions add together, or they may be opposed to each other and so their dispersions partially cancel. The range of dispersions given by these options varies from 200 to 800 nm mm^{-1} in the red to 47 to 180 nm mm^{-1} in the blue
Spectral resolution	ranges from 37 to 150 in the red and from 40 to 97 in the blue for the various options

on Schmidt cameras, however. The wide field of these instruments means that upwards of a hundred thousand spectra can be obtained in a single exposure, more than offsetting the low efficiency of photography as a detector. The parameters of the objective prisms for the 1.2 m UK Schmidt camera are listed in table 10.10. This camera can also undertake multi-object spectroscopy with images being fed to a conventional laboratory-based spectroscope by fibre optics. This instrument, known as FLAIR (Fibre-Linked Array-Image Reformatter), suffers from a similar constraint to the LDSS (see above) in that an accurately machined plate has to be made for every observation. The fibre optic feeds are mounted onto this plate and it then has to be curved to match the focal surface of the Schmidt camera.

10.14   THE FUTURE

In the absence of some radically new discovery, the future design of spectroscopes seems likely to be a slow evolution towards higher efficiency for the existing types of instrument, and for the use of higher spectral resolutions, at least for some applications. Individual optical components and detectors can already achieve 50% to 80% efficiencies, and so there is only limited scope for improvement there. The overall throughputs for telescope–spectroscope–detector combinations, however, can still be very low. It is here therefore that improvements are likely to arise, especially via a reduction in the number of optical components by the increased use of off-axis surfaces for mirrors and aspherical surfaces for lenses. Variable refractive indices for materials, produced by, for example, the diffusion of silver into glass, are another possible way in which the numbers of components may be reduced. Greater overall efficiency will also be achieved by extending and improving the current trend to multi-object spectroscopy.

Part 3

Spectroscopy of Astronomical Sources

11

Spectral Classification

11.1 SPECTRAL CLASS

11.1.1 Core classification

Brief mention of the classification of stars on the basis of the appearance of their spectra has been made in chapter 1, and the historical roots of the rather awkward system that has been evolved discussed there. Here, therefore, we look at the process of spectral classification itself, and the related process of luminosity classification.

The spectral class of a star is based upon the appearance of the lines in its spectrum over the blue–green region, from about 380 nm to 500 nm, and is a classification in terms of the temperature of the stellar surface layers. There are seven major groups in the core of the classification system, labelled, as already mentioned in chapter 1, with the letters

$$O \quad B \quad A \quad F \quad G \quad K \quad M$$

in order of decreasing temperature. The mnemonic 'Oh Be A Fine Girl/Guy Kiss Me' may be found useful for remembering the sequence. Each of these major groups is subdivided into ten by the addition of an arabic numeral between 0 and 9. The Sun, with a surface temperature of about 5700 K is thus classified as a G2 star. The hottest stars so far discovered that are classified in this system are O4 stars at about 40 000 K, and the coolest are M8 stars at about 2500 K. Hotter stars exist than O4; white dwarfs in particular may have temperatures up to 100 000 K, but the pressures at their surfaces are so high that few if any spectral lines may be detected (see later discussion and chapter 13), and so they cannot be included within a classification system that is based upon spectral lines. Cooler stars than M8 also almost certainly exist and have been given the rather misleading name of 'Brown Dwarfs'. They are transitional objects between the smallest of the true stars and the largest of the planets. Such objects are the subject of much current observational effort, but at the time of writing

Table 11.1   The full range of spectral classes in current use.

O4	B0	A0	F0	G0	K0	M0
O5	B0.5	A2	F2	G2	K2	M1
O6	B1	A3	F3	G5	K3	M2
O7	B2	A5	F5	G8	K4	M3
O8	B3	A7	F7		K5	M4
O9	B5		F8			M7
O9.5	B7		F9			M8
	B8					M9
	B9.5					

no unequivocal identification has been made. The relationship between effective surface temperature and spectral class is shown in figure 11.1, for main sequence stars.

More recently the original decimal subdivision of the core classes has been adapted to provide a smoother variation with temperature by adding and deleting some of the classes. Classes have been added between O9 and B0, B0 and B1, and B9 and A0 to give O9.5, B0.5 and B9.5. Classes B4, B6, B9, A1, A4, A6, A8, A9, F1, F4, F6; G1, G3, G4, G6, G7, G9, K1, M5 and M6 have been deleted (K6, K7, K8, K9 were not much used from the beginning); see table 11.1. Some of these classes, however, continue in use in the literature today because the classification in many star catalogues pre-dates the more recent changes.

The appearance of stellar spectra is dominated by a few strong lines which change slowly over the spectral classes (figure 11.2). The principal lines in each class are:

O   Mostly highly ionized silicon, nitrogen etc, plus ionized helium (the defining characteristic)

B   Neutral helium (the defining characteristic), no ionized helium, lower stages of ionization of silicon and nitrogen, Balmer lines strengthen

A   Balmer lines peak at A0, lines from singly ionized calcium and metals appear

F   Balmer and ionized metal lines weaken, lines from neutral metals strengthen

G   Balmer lines continue to weaken, ionized calcium lines peak in intensity, lines of neutral metals continue to increase in strength

K   Many lines due to neutral metals, molecular bands (TiO) start to appear

M   TiO bands dominate.

The reason why the lines in a spectrum change with temperature is largely due to the balance between the ionization and the excitation of the atoms composing the outer layers of the star. We have already seen that the population of different excited levels is temperature-dependent, and is given in thermodynamic equilibrium by Boltzmann's formula (equation (4.5)). A useful approximation for much work on stellar atmospheres is that of Local Thermodynamic Equilibrium (LTE). In this approximation, the properties of the material are assumed to be characterized by a single temperature, and the

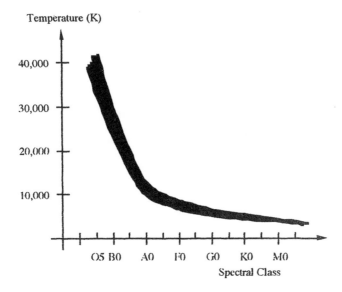

Figure 11.1 The relationship between the effective surface temperature of a star and its spectral class.

properties of the radiation field also characterized by a single, but *different*, temperature. If LTE may be assumed, as is often the case except for the most critical work, then Boltzmann's formula may be used to give the relative populations of the different excited levels of the atoms in the surface layers of a star. The levels of ionization are similarly given in LTE by the Saha equation:

$$\frac{N_{i+1}}{N_i} = \frac{2U_{i+1}}{N_e U_i}\left(\frac{2\pi m_e kT}{h^2}\right)^{3/2} e^{-\chi_i/kT} \tag{11.1}$$

where N_i is the number density of an atom in the ith stage of ionization, N_{i+1} is the number density of an atom in the $(i+1)$th stage of ionization, N_e is the free electron number density and χ_i is the ionization potential for the ith ion. U_i and U_{i+1} are the total statistical weights for the ith and $(i+1)$th ions. They are called Partition Functions, and are obtained from the individual statistical weights of the energy levels by the equation (often only the first term, or the first few terms, need to be calculated)

$$U_i = g_{i0} + g_{i1}e^{-E_{i1}/kT} + g_{i2}e^{-E_{i2}/kT} + \ldots \tag{11.2}$$

where g_{ix} is the statistical weight of the xth level of the ith ion and E_{ix} is the excitation potential of the xth level of the ith ion, and they may also be seen to depend upon the temperature. Other things being equal, the strength of an absorption line depends upon the number of atoms available to produce it. Lines from the ground state (Resonance lines) of an atom are therefore usually

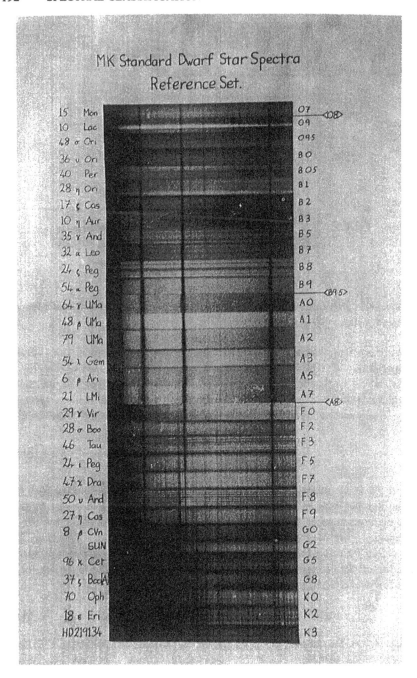

Figure 11.2   The spectral sequence (photograph reproduced courtesy of Mrs H Reeder, University of Hertfordshire Observatory).

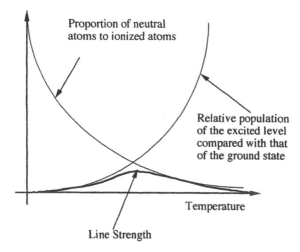

Figure 11.3 Variation of the line strength for a line from an excited level (schematic).

strongest at low temperatures and decrease in strength as the temperature rises, because atoms in the ground state are lost by excitation to higher levels and by ionization. Lines arising from an excited level by contrast will be weak at low temperatures because few atoms will be excited to that level, and will become stronger as the temperature rises and the originating level becomes more highly populated. Ionization, however, will eventually reduce the line strength at very high temperatures by removing the atoms to the next stage of ionization. Such lines, of which the hydrogen Balmer series is a good example, therefore show an increase in strength as the temperature rises, followed by a decrease (figures 11.3 and 11.4). Lines from ions follow similar patterns except that the temperature must be sufficiently high to produce a significant number of the ions before any of the lines may appear.

Measurement of the absolute strengths of spectral lines is possible and could therefore be used to provide spectral classification. But obtaining even moderately accurate absolute measurements is a lengthy process, and therefore spectral classification is actually based upon relative intensities. If lines from two different atoms or ions are carefully chosen, then one of the lines may be increasing in strength as the temperature rises, while the other is decreasing. The relative strength of one such line with respect to that of the other is therefore a sensitive measure of the temperature over the range where the two lines' strengths vary in opposite ways. So sensitive is the variation of such a line strength ratio to the temperature that it can be estimated by eye with a little practice, quite accurately enough to classify the stars. The most sensitive such ratios are not usually to be found amongst the lines which dominate the spectrum, but occur for the weaker lines, and the principal line pairs used for classification of main sequence stars are listed in table 11.2.

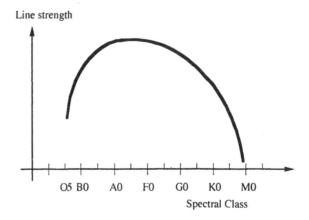

Figure 11.4 Variation of the strength of the Balmer series in stellar spectra.

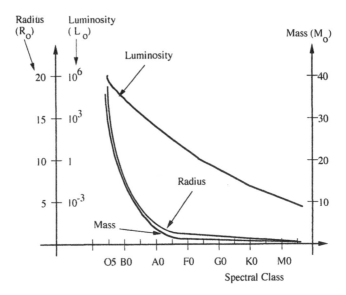

Figure 11.5 The variation of stellar size, mass and luminosity for main sequence stars with spectral class.

The core spectral class thus provides a straightforward way for classifying the main sequence stars, with relatively little effort. A medium dispersion spectrum in the blue–green region is all that is required and the class is estimated from that by eye. Once the core spectral class of a main sequence star is known then its temperature (figure 11.1) and many other factors such as its size, luminosity etc (figure 11.5) may be determined with reasonable accuracy.

Table 11.2 Criteria for spectral classification.

Spectral classes	Line pairs
O4–B0	He II 454.2 / He I 447.1
O7–O9	He II 420.0 / He I 402.6+He I 414.4
O7–B0	He II 468.6 / He I 492.2
B0–B0.5	Si IV 408.9+Si IV 411.6 / He I 412.1 and He II 468.6 / He I 471.3
B0–B1	C III 406.8–407.0 / He I 400.9 and C III 464.7–465.1 / He I 471.3
B0–B2	Si III 455.2 / Si IV 408.9
B1–B5	C II 426.7 / Mg II 448.1
B2–B8	Mg II 448.1 / He I 447.1
B3	Si II 412.8–413.0 / He I 412.1
B5–B7	C II 426.7 / Mg II 448.1, Si II 412.8–413.0 / He I 414.4
B5–A0	Si II 412.8–413.1 / He I 414.4+He I 402.6, Mg II 448.1 / He I 447.1 and Balmer line profiles
B9.5–A5	Mn I 403.0–403.4 / Fe I 427.1
A0–A7	Mg II 448.1 / Fe I 438.5
A0–F0	Ca II 393.4 / Ca II 396.8+H I 397.0
F0–G0	Fe I 404.6 or Ca I 422.7 / H I 410.2 or H I 434.0, Mn I 403.0–403.4 / Si II 412.8–413.2 and CH 430.0 / Fe I 438.5
G0–K0	Fe I 404.6 or Fe I 414.4 or Ca I 422.7 / H I 410.2, Fe I 438.4 / H I 434.0, Fe I 492.1 / H I 481.6
K0–M0	Ti I 399.0+Ti I 399.9 / Fe I 400.5, Ti I 400.9–401.0 / Fe I 400.5, Fe I 414.4 / H I 410.2, Ca I 422.7 / Fe I 425.1, Cr I–Fe I 429.0 / Fe I 432.6 and Cr I 425.4 or 427.4 / Fe I 425.0 or Fe I 426.0 or Fe I 427.1
M0–M8	TiO band strength
M4–M8	CaOH 550–560 band
M7–M8	VO bands

Data obtained from *An Atlas of Representative Stellar Spectra* by Y Yamashita, K Nariai and Y Norimoto, University of Tokyo Press, 1977 and *The Classification of Stars* by C Jaschek and M Jaschek, Cambridge University Press, 1990.

11.1.2 Additions to the core spectral class

Special features in the spectra of stars otherwise classified within the core spectral classes may be indicated by lower case letters placed before or after the core spectral classification. There is some overlap here with the luminosity classification (see below). The main such symbols in current use are listed in table 11.3.

11.2 THE NON-CORE CLASSES

There are six other major classes of spectra after the seven within the core. They are given the symbols

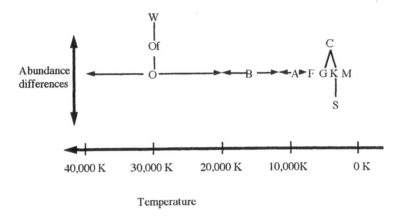

Figure 11.6 The spectral classification system.

R N S W P Q

with R and N sometimes combined into class C. The usage of these classes varies considerably. Classes P (nebulae) and Q (novae and supernovae) are almost never encountered today, and will not be considered further here. Classes R, Ṅ (or C) and S find reasonably widespread use, while the use of Class W (Wolf–Rayet stars) is almost universal. With the exception of classes P and Q, these groups have significant differences in elemental abundance from core class stars.

Wolf–Rayet stars (class W) have spectra with extremely wide emission lines, sometimes with absorption components on the short wavelength edges (a P Cygni line profile, see chapter 13). Their surface temperatures place them in the same range as the O-type stars, but hydrogen appears to be in low abundance. Instead, the helium lines are very intense. There are two subclasses, the WN stars, with strong nitrogen emission lines in addition to the helium lines, and the WC stars, with strong carbon and oxygen lines as well as helium lines. Each of these subclasses is divided, like the core classes, by a decimal subdivision dependent upon temperature. Thus the brightest of the Wolf–Rayet stars, γ^2 Velorum is WC8 (actually it is classified as WC8 + O7v, because it is a binary star). As already remarked, the Of stars are intermediate in composition between the Wolf–Rayet stars and the O-type stars. They are defined by the presence of emission lines of ionized helium at 468.6 nm and of doubly ionized nitrogen at 463.0–463.4 nm.

The R and N stars correspond in terms of temperature to G and K in the core classes. They are variable giant stars with strong molecular bands due to carbon compounds, especially cyanogen (CN) and the Swan bands of molecular carbon (C_2, figure 5.18), in their spectra. Their alternative designation as C-type stars thus derives from this excess of carbon. They also have other abundance peculiarities. The S-type stars are similar in temperature to the K and N stars.

Table 11.3 Special characteristics of the core spectral classes.

Suffix	
e	Emission lines present in the spectrum. If these include the hydrogen lines, then the last such to be in emission may (sometimes) be indicated by its Greek letter; e.g. B8eγ, if Hγ were the last line observed to have an emission component when going down the Balmer series from Hα to shorter wavelengths
f	Used only for stars intermediate in nature between the O-type stars and the Wolf–Rayet stars (see below), which are labelled as Of stars. The symbol is a remnant of an early variation on the core classification system whereby for the O and M stars, the decimal subdivision was indicated by lower case letters (Oa, Ob, Oc etc)
k	Interstellar lines in the spectrum
m	Lines from the metals present unusually strongly
n	Very broad (i.e. Nebulous) lines. When the linewidths are due to pressure broadening, then such stars may also be subdwarfs or white dwarfs (sd, w, wd, D, see below, see also luminosity classification)
p, pec	A spectrum with a peculiarity not covered by the other special symbols (e.g. Ap stars which amongst other oddities have very high abundances of rare-earth elements)
s	Unusually narrow (Sharp) lines, though not usually as narrow as the lines in supergiants
v	A variable spectrum

Prefix (these symbols are encountered much less frequently than those following the core classification)	
c, sg	Very narrow spectral lines (narrower usually than 's' above). The star is a supergiant (luminosity class O, I, Ia, Ib, see below)
d	Dwarf or main sequence star (luminosity class V, see below)
g	Giant (luminosity class III, see below)
sd	Subdwarf (luminosity class VI, see below and 'n' above)
w, wd, D	White dwarf (luminosity class VII, see below and 'n' above)

Examples of the use of these symbols	
Betelgeuse	M2ev
Sirius B	wA5
Mira	M6ev
κ Ori	cB0.5e
Pollux	gK0
εUMa	A0p

They have very strong ZrO bands in their spectra, and the more usual TiO bands are weak or absent. The subdivision of these classes has two components. The

Table 11.4 Luminosity classes.

Designation	Star
I, Ia, Iab, Ib	Supergiant (class 0 is sometimes used for truly exceptionally large and bright individual stars like P Cyg)
II	Bright giants
II-III, IIIa, IIIab, IIIb, III-IV	Giants
IV	Subgiants
V	Main sequence/dwarf stars
VI	Subdwarfs
VII	White dwarfs

Examples	
Sun	G2 V
Sirius A	A1 V
Sirius B	A5 VII
Betelgeuse	M2ev I

first decimal figure is based upon temperature in a similar manner to the variation in the core classes. The second figure indicates the degree of intensity of the defining peculiarities. This last classification is highly complex and details are left for the reader to find in specialist literature. As examples of these classes we have

$$19\,\text{Psc} \quad \text{C7}, 2$$
$$\text{HD}\,53432 \quad \text{C4}, 5$$
$$\text{S UMa} \quad \text{S0.5}, 9$$
$$\chi\,\text{Cyg} \quad \text{S7}, 2$$

Pictorially, we may represent the main classes currently in use today as shown in figure 11.6.

11.3   THE LUMINOSITY CLASS

Amongst the many factors that can broaden spectral lines (chapter 13), one of the most widespread is pressure. As the pressure in a star's outer layers increases, so more and more atoms and electrons may pass close to an atom or ion during the time that it is emitting or absorbing a photon. Such close passages disturb the atom and change the photon's energy. The aggregate effect of this on many emitting or absorbing atoms is that the photons forming the line have a wider distribution in wavelength than normal, i.e. the line is broader than normal. The broadening increases as the pressure increases.

Table 11.5 Principal line pairs usable for luminosity classification.

Useful range of temperature class	Line pairs usable for luminosity classification
O6–O8	He II 468.6 and N III 463.4–464.2 emission / N IV 347.9–348.5 absorption
O9–O9.5	Si IV 408.9 /C III 406.8–407.0, Si IV 411.6/He I 412.1, C III 464.9/He II 468.6 and He II 468.6/C III 464.7–465.1+He I 471.3
B0–B1	Si IV 411.9/He I 402.6, 414.4; 400.9, Si IV 411.6/He I 412.1 and Si III 455.3/He I 471.3
B1	Si IV 408.9, 411.6/He I 412.1 (luminosity classes I–III only), N II 399.5/He I 400.9, O II 441.5–441.7/He I 438.8, Si III 455.3/He I 471.3, 438.8 and O II-C III blend at 465.0/ He I 471.3, 438.8
B2	O II and Si III lines do not appear in classes V and VI, but are strong in class I: He I 412.1/He I 414.4, He I 396.5/He I 400.9, N II 395.5/He I 400.9, O II 441.5–441.7/He I 438.8, Si III 455.3/He I 471.3, 438.8, N II 463.1/He I 471.3
B3	C II 391.9–392.1/He I 392.7, N II 395.5/He I 400.9, He I 396.5/He I 400.9, He I 412.1/He I 414.4 and C II 426.7/He I 438.8, 414.4
B5–B7	He I 396.5/He I 400.9, He I 412.1/He I 414.4, Mg II 448.1/He I 447.1 and C II 426.7/He I 438.8 and the Balmer line profiles. N II 399.5/He I 400.9 can be used to separate out the subclasses of luminosity class I (I, Ia, Iab, Ib)
B7–B8	The wings of the Balmer lines. C II 426.7/Fe II 423.3 may be used for classes I and II. He I 396.5/He I 400.9 and He I 412.1/He I 414.4 can be used to separate out the subclasses of luminosity class I (I, Ia, Iab, Ib)

Table continued overleaf

Now most normal stellar masses lie in the range from a quarter of a solar mass to 20 solar masses, but their radii range from 0.01 to 1 000 solar radii. Their average densities must therefore range from about 10^{-8} to 10^{6} times that of the Sun, with the smaller the star, the higher its density. Now although mean density is not the only factor governing pressure in the outer layers of a star, it is the major influence. Hence pressure in the outer layers of a star is also generally greater, the smaller the star. The lines in a spectrum, all other things being equal, thus increase in width as the size of the star decreases. Finally, the physically larger stars are also usually brighter than the smaller ones, so that as the spectral line widths are observed to increase, so the star will usually be found to be fainter. This is the basis of the second component of spectral classification, the Luminosity Class.

The luminosity class is added as a Roman numeral from I to VII, after the main temperature spectral class. Later work has added several intermediate

Table 11.5 Continued.

Useful range of temperature class	Line pairs usable for luminosity classification
A0–A2	Si II 385.6/Ca II 393.4, Fe II 423.3/Mg II 448.1 and Si II 412.8/Mg II 448.1. The Balmer line wings can be used between classes III and I
A2–A3	The Balmer line wings plus Fe II–Ti II blend at 402.5/Fe I 404.6, Fe II 423.3/Fe II 422.7 and Fe II 441.7 and Fe I, II 438.3–438.5/Mg II 448.1
A3–A7	The Balmer line wings plus Ti II 401.2/Fe I 400.5, Fe II–Ti II blend at 402.5/Fe I 400.5, 404.6, Fe II–Ti II blend at 417.3+Fe II 417.9/Fe I 414.4, Fe II 423.3/Ca I 422.7 and Ca I 441.7 and Fe II, Ti II 441.7/Mg II 448.1
A9–F1	Y II 398.3/Fe I 400.5, Fe II–Cr II 400.2 blend/Fe I 400.5, Fe II–Ti II blend at 402.5/Fe I 400.5, 404.6, Sr II 407.8/Fe I 404.6, Fe II 417.9/Fe I 414.4, Fe II–Ti II blend at 417.3 /Ca I 422.7 and Sr II 421.6/Ca I 422.7
F2–F5	Y II 398.3/Fe I 400.5, Fe II–Cr II 400.2 blend/Fe I 400.5, Fe II–Ti II blend at 402.5/Fe I 400.5, 404.6, Sr II 407.8/Fe I 404.6, Fe II 417.9/Fe I 414.4, Fe II–Ti II blend at 417.3 /Ca I 422.7+Fe I 414.4, Sr II 421.6/Fe I 414.4, Fe I 441.7 and Ti II 444.4/Mg II 448.1 and Ba II 455.4/Mg II 448.1
F7–F9	Y II 398.3/Fe I 400.5, Fe II–Ti II blend at 402.5/Fe I 400.5, Sr II 407.8/H I 410.2+Fe I 404.6, Fe II 412.9, 417.9/Fe I 414.4, Sr II 421.6+Sc II 424.7/Fe I 414.4 and Ba II 455.4/Mg II 448.1
G0–G2	Y II 398.3/Fe I 400.5, Sr II 407.8/H I 410.2+Fe I 404.6, 406.4, Fe II 417.9/Fe II 417.3 and Sr II 421.6/Fe I 414.4
G5–G8	Sr II 407.8/Ca I 422.7+Fe I 404.6, 406.4, Sr II 421.6/Fe I 414.4, 427.2+Ca I 422.6 and Ti II 440.0, 440.8/Fe I 440.5
G8–K0	CN discontinuity in the continuum at 421.6, plus Sr II 407.8/Fe I 404.6, 406.4, 407.2, Sr II 421.6/Fe I 414.4, 427.2, Ti II 440.0, 440.8/Fe I 440.5 and Fe I 444.4/Fe I 440.5
K2–K4	CN discontinuity in the continuum at 421.6, plus Sr II 407.8/Fe I 404.6, 406.4, 407.2, H I 401.2/Fe I 406.4, 407.2, 414.4, Sr II 421.6/Fe I 414.4, 427.2 and Ti II 440.0, 440.8/Fe I 440.5
K5–M0	Sr II 407.8/Fe I 406.4, 407.2, H I 401.2/Fe I 406.4, 407.2, 414.4 and Sr II 421.6/Fe I 414.4+Ti I 418.6
M0–M1	Sr II 407.8/Fe I 404.6, 406.4, H I 401.2/Fe I 414.4, Sr II–Fe I blend at 421.6/Fe I 414.4 and Fe I 437.5, 438.9/Fe I 438.4
M1–M4	Sr II 407.8/Fe I 404.6, 406.4, 407.2, 426.3, H I 401.2/Fe I 404.6, 414.4, Sr II–Fe I blend at 421.6/Fe I 414.4, 425.1 and Fe I 437.5, 438.9/Fe I 438.4

Data obtained from *An Atlas of Representative Stellar Spectra* by Y Yamashita, K Nariai and Y Norimoto, University of Tokyo Press, 1977 and *The Classification of Stars* by C Jaschek and M Jaschek, Cambridge University Press, 1990.

classes. The classes are listed in table 11.4. Like the temperature classes, the luminosity classification is based upon eye estimates of the relative intensities

Table 11.6 Subclassification of the white dwarfs.

Class	Apparent element abundance
DA	Hydrogen rich
DB	Helium rich
DC (or BC)	No detectable lines (i.e. a continuum spectrum)
DO	Ionized helium lines strong
DQ	Carbon rich
DZ	Lines of the metallic elements only

of pairs of lines, and the principal ones in use are listed in table 11.5.

11.3.1 White Dwarfs

Pressure broadening of spectral lines reaches its maximum in neutron stars. However, so few of these are detectable at optical wavelengths that they are not included in the classification system. White dwarfs are therefore the densest stars given a luminosity class. The lines are so broad in white dwarf spectra, however, that they are difficult to include in the temperature class in the normal way. Indeed, sometimes the lines merge together and we get a spectrum which is close to that of a pure continuum. The white dwarfs are therefore classified on the basis of their (apparent) abundances, as shown in table 11.6.

12

Radial Velocities

12.1 INTRODUCTION

The component of the relative velocity between the Earth and an astronomical object along the line of sight is called the radial velocity of that object. Since the wavelengths of lines are changed by such a velocity, as described by the Doppler shift, it is a relatively easy parameter to measure, and the earliest determinations date back to Huggins' visual spectroscopy (chapter 1). The Doppler formula gives the radial velocity from the change in wavelength (or frequency), with the convention that velocities away from the Earth are positive:

$$\frac{\Delta\lambda}{\lambda} = -\frac{\Delta\nu}{\nu} = \frac{v}{c} \qquad (12.1)$$

where λ and ν are the rest wavelength and frequency of the line respectively, $\Delta\lambda$ is the wavelength change ($\Delta\lambda = \lambda_0 - \lambda$), $\Delta\nu$ is the frequency change ($\Delta\nu = \nu_0 - \nu$), and λ_0 and ν_0 are the observed wavelength and frequency of the line respectively, i.e.

$$v = \frac{\lambda_0 - \lambda}{\lambda}c = \frac{\nu - \nu_0}{\nu}c. \qquad (12.2)$$

Equations (12.1) and (12.2) are applicable to velocities below about $10\,000$ km s^{-1}. As the velocities become an appreciable fraction of the speed of light, however, the relativistic version of the Doppler formula must be used:

$$\frac{\Delta\lambda}{\lambda} = -\frac{\Delta\nu}{\nu} = \frac{1 + (v\cos\theta/c)}{\sqrt{1 - (v^2/c^2)}} = z \qquad (12.3)$$

where θ is the angle between the line of sight and the relative velocity vector, and z is the symbol conventionally used for redshift at high recessional velocities. We thus have for pure radial velocity ($\theta = 0°$):

$$v = \left(\frac{\lambda_0^2 - \lambda^2}{\lambda_0^2 + \lambda^2}\right)c = \left(\frac{(2\nu - \nu_0)^2 - \nu^2}{(2\nu - \nu_0)^2 + \nu^2}\right)c. \qquad (12.4)$$

Normally the Doppler effect only allows the radial velocity to be determined, but at relativistic velocities, the $\cos\theta$ term in equation (12.3) means that there is a wavelength change even for velocities directly across the line of sight. The transverse Doppler effect is obtained by setting $\theta = 90°$, and is given by

$$z_T = \frac{\lambda_0 - \lambda}{\lambda} = \frac{\nu - \nu_0}{\nu} = \frac{1}{\sqrt{1 - (v^2/c^2)}} - 1. \tag{12.5}$$

As can be seen from equation (12.5), the transverse shift is always to longer wavelengths. The transverse Doppler effect is rarely encountered; the peculiar object SS433 is the best known example, with its jets moving nearly across the line of sight at about a quarter of the velocity of light.

Heliocentric radial velocities are usually quoted. That is, the object's velocity with the line of sight component of the Earth's orbital motion removed. The heliocentric velocity (v_H) is related to the measured or geocentric velocity (v_G) by

$$v_H = v_G + v_E \left[\sin\delta_A \sin\delta_O + \cos\delta_A \cos\delta_O \cos(\alpha_O - \alpha_A)\right] \tag{12.6}$$

where v_E is the Earth's orbital velocity (29.78 km s^{-1} on average) α_O and δ_O are the right ascension and declination of the object, and α_A and δ_A are the right ascension and declination of the apex of the Earth's orbital motion at the time of the observation. They are given by

$$\delta_A = \sin^{-1}(\cot\delta_s \sin\alpha_s \sin\varepsilon) \tag{12.7}$$

$$\alpha_A = \alpha_s - \cos^{-1}(-\tan\delta_A \tan\delta_s) \tag{12.8}$$

where ε is the obliquity of the ecliptic (23° 27') and α_s and δ_s are the right ascension and declination of the Sun at the time of the observation.

12.2 TRADITIONAL APPROACH TO RADIAL VELOCITY DETERMINATION

The traditional approach to measuring astronomical radial velocities is from a photographic spectrogram which has a wavelength comparison spectrum (figure 9.6). As a rough rule of thumb, the radial velocity may be found to the same level of accuracy in km s^{-1} as the dispersion of the spectrum in Å mm^{-1} (i.e. ± 10 km s^{-1} from a spectrum with a dispersion of 10 Å mm^{-1}, or 1 nm mm^{-1}). The procedure is time-consuming (and very boring!). The steps in the process are as follows.

1. Identify the lines in the comparison spectrum and the object's spectrum (see below). Suitable tables of wavelengths are listed in appendix C. The comparison wavelengths, however, will normally be known from previous work.

The more intense lines in the object's spectrum may usually be identified without difficulty unless the radial velocity is very large indeed (as for example with Quasars). The fainter lines may only be identifiable once an initial estimate of the radial velocity has been obtained and their corrected wavelengths found. Since the fainter lines will usually be narrower, and can be measured with a higher precision than the stronger lines, it is often necessary to approach the radial velocity by successive approximations.

2.   Place the spectrum onto a measuring machine (figure 9.7). These machines take a variety of forms, but are all essentially high precision travelling microscopes which enable the positions of features in the spectrum to be measured to an accuracy of about one micron. Usually the microscope is fixed and the spectrum placed on a table which can be moved underneath it. The motion of the table is controlled by a precision screw thread from which its position may be determined by a Vernier indicator. The spectrum is aligned with the direction of motion of the table, and the measuring cross wires in the microscope eyepiece aligned parallel with the spectral lines (which are not always precisely perpendicular to the length of the spectrum).

3. Measure all the desired lines in both the comparison spectrum and the object's spectrum in a single pass down the spectrum, taking care to keep the screw moving in a single direction to avoid problems with backlash.

4. Turn the spectrum through 180°, re-align, and repeat the measurements for all the lines.

5. Return the spectrum to its original orientation and repeat steps 2, 3 and 4 as required.

6. Calling the measurements (or their averages) taken with the spectrum in its first orientation, x, and those with it reversed, y, a check on the accuracy of the results may be obtained by adding the x and y for each line together. The result of this addition should be a constant, representing the separation of the zero points for the measurements in the two orientations (figure 12.1). The scatter of the results of this addition enables an estimate of the accuracy of the measurements to be obtained. Any individual results which diverge significantly from the average are probably due to one or more mis-measurements of the line involved, and the data for that line have to be discarded.

7. The overall average of both the x and y measures may be obtained by subtracting the one from the other (there is no need to divide by the total number of measurements, since this just represents a scaling factor). The value of $x - y$ then represents the position of the line with respect to a point midway between the zeros for the x and y measurements (figure 12.1), which we may call m.

8. The first approximation to the relationship between position along the spectrum and wavelength may now be found. For a grating-based spectrum, the relation is approximately linear (equation (8.14)), and so we may use an equation of the form

$$\lambda = A + Bm \tag{12.9}$$

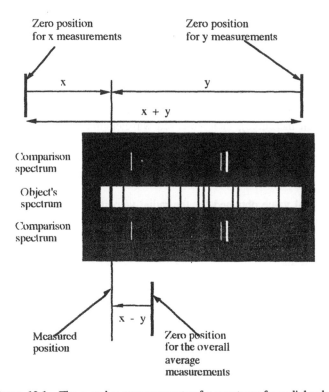

Figure 12.1   The x and y measurements of a spectrum for radial velocity.

where A and B are constants. For a prism-based spectrum, the relationship is nonlinear (equation (8.31)), and is more accurately represented by

$$\lambda = A + \frac{B}{m - C}$$ (12.10)

where A, B and C are constants. For a grating spectrum therefore, two comparison lines are selected about a third of the way in from each end of the spectrum. Using their measurements and known wavelengths, equation (12.9) is solved simultaneously to obtain values of A and B. For a prism-based spectrum three comparison lines are chosen, and equation (12.10) solved for A, B and C.

   9. Equations (12.9) and (12.10) are only approximations. A more accurate relationship between measurement and wavelength has therefore to be found by constructing a correction curve. The correction curve is obtained by using the measurements for all the comparison lines, and determining measured wavelengths for them using either equation (12.9) or (12.10) as appropriate. This measured wavelength is then subtracted from the actual wavelength for each comparison line, and the difference (usually small) plotted against the actual wavelength to produce the correction curve.

10. The observed wavelengths for the lines in the object's spectrum may now be determined. The initial approximation is obtained using the value for m for each line substituted into equation (12.9) or (12.10). These are then corrected for the approximate nature of the formulae using the correction curve.

11. The radial velocity of the object may now be found via the Doppler formula and the rest wavelengths for the lines in the object's spectrum.

12. If some lines in the object's spectrum could not previously be identified (but have been measured), their observed wavelengths may now be corrected for the radial velocity shift. The more precise values for their rest wavelengths may then allow identifications (see later) to be made. If need be, a second approximation to the radial velocity may then be found by returning to the previous stage.

13. The final value obtained for the radial velocity is then corrected for the Earth's motion (equation (12.6)) to give the heliocentric radial velocity.

12.2.1 Shortcuts

The traditional approach to determining radial velocities outlined above is clearly lengthy, especially if the measurement stages are repeated a number of times. Various approaches have therefore been tried to reduce the labour involved. However, for the highest precision work there is not much that can be done, though the strain on the operator can be reduced by using a TV camera on the microscope and working from the display on the monitor. A rather more advanced approach scans a short section of the spectrum and displays it both as a direct tracing and a reversed tracing. The measurement is made by superimposing the two tracings of a line. This approach also reduces the problem of measuring asymmetrical lines since it is easier to judge when two such profiles are aligned than to try to set a cross wire on the 'centre of gravity' of a single line. Electronic read-outs for the spectral line positions can be fed directly into a small computer at the press of a button by the operator, speeding up the whole process and eliminating mis-readings of the Vernier settings. Once the data are in the computer, then stages 6–13 can all be processed there as well.

If the radial velocity is required to less accuracy, perhaps to provide approximate rest wavelengths for unidentified lines, then several short cuts are possible. Firstly, simply fewer lines may be measured, with only one or two measurement sequences, for both the comparison spectrum and the object's spectrum. Secondly, if two comparison lines are close to a line from the object, then its wavelength may be found by straightforward interpolation. Thirdly, tracings of the spectra may be made and the wavelengths measured from them. For this to work, either the microdensitometer must be of a design which allows several tracings to be superimposed, without slippage, so that the comparison and object spectra may be traced separately, or the microdensitometer slit must be lengthened so that both the comparison lines and the object's lines are covered in a single tracing (figure 12.2). This last approach reduces the contrast considerably because of the greatly increased background coming through the long slit, and so fainter lines may not be distinguishable.

Some measuring machines scan the entire image and reduce it to a machine-readable form. In such cases, the data may be processed as discussed for directly produced machine-readable spectra below. Most such measuring machines (COSMOS, APM etc), however, are intended for scanning large plates from Schmidt cameras rather than the small plates commonly produced by photographic spectroscopes.

12.3 OBJECTIVE PRISM SPECTRA

The objective prism spectroscope (chapter 8) is highly efficient, both because of its high throughput, and because of the large number of objects simultaneously observed. Accurate radial velocities, however, cannot be obtained from objective prism spectra because they are usually of low dispersion, and it is not possible to provide a good comparison spectrum. One way around the problem, which will provide low accuracy data, is to take two exposures on the same plate, with the objective prism rotated through 180° between the two exposures (figure 8.24). If two or three or more stars with known radial velocities are on the plate, then the separation of a line in the two spectra may be calibrated in terms of velocity, and the velocities of the other stars determined.

Another approach is to use lines from terrestrial sources. The Earth's atmosphere produces absorption lines in the red and infrared parts of the spectrum; unfortunately this is the lowest dispersion part of an objective prism spectrum. A filter containing neodymium chloride has a single sharp absorption feature at 427 nm. Though the wavelength of the feature is temperature dependent, these filters have been used with some success.

For very low precision work, such as quasar identification, where velocities of $100\,000$ km s^{-1} or more need only be estimated to $\pm 10\%$ in order to provide candidates for more detailed study, then the cut-off in sensitivity of the photographic emulsion can be used. For some emulsions this can be quite sharp, giving a wavelength marker with an accuracy of 10 or 20 nm.

Treanor's method does not use an objective prism, but it is sufficiently closely related in principle to be dealt with here. The wavelength marker in the spectrum is a direct image of the object. Treanor's prism is a direct-vision prism placed in the collimated beam of light from a telescope. The prism is physically smaller than the beam of light, so that the spectrum produced by the prism has a direct image superimposed upon it from the light passing around the prism (figure 12.3). Like reversing an objective prism, this system also requires some stars with known radial velocities to be on the image in order to provide a calibration.

12.4   MACHINE-READABLE SPECTRA

Many spectroscopes (chapter 10) now use detectors such as CCDs, IPCS, SEC Vidicon TVs etc, which have outputs fed directly into a computer. The principle of the determination of radial velocities from such spectra is the same as that for the traditional approach (above), but most of the labour is now undertaken by the computer. Line identification for the object's spectrum still usually has to be undertaken 'by hand' unless numerous very similar spectra are being processed. The comparison lines, however, will normally be the same from one spectrum to the next, and so their data only need to be provided once. The positions of lines are determined by fitting an appropriately shaped profile, or by the determination of the position of the 'centre of gravity'. Steps 8 and 9 of the traditional approach may simply be transferred to the computer, or more accurate formulae than equations (12.9) and (12.10) used from the start. Most modern spectroscopes will have appropriate software already written and readily available, so that the observer is hardly aware of the details of the process, but just obtains the radial velocities as his or her 'raw' data directly.

12.5   GRIFFIN'S METHOD

A ingenious direct reading radial velocity apparatus is due to Griffin. This employs a spectroscope of fairly conventional design, but places a previously obtained negative image of a spectrum at the focus. The transmitted light is then re-imaged onto a point source detector, and the total output measured. The negative spectrum must be from a similar type of object to that being observed (e.g. an A5 stellar spectrum could be used for stars say in the range A3 to A7). The negative image is moved along the real image, and when the two coincide there is a sharp drop in the output from the detector. This is because the brightest parts of the real spectrum will then coincide with the darkest parts of the negative, and the darkest parts of the real spectrum with the most transparent parts of the negative, whereas in other positions some of the bright parts of the real image will come through the transparent parts of the negative. The displacement of the negative can then be calibrated to read the object's velocity directly. The apparatus is particularly suited to the study of spectroscopic binary stars, where many radial velocity measurements of the same object are required.

12.6   LINE IDENTIFICATION

Most of the time, the identity of the main lines in a spectrum will be known from previous work by other astronomers (figure 11.2, for example). The fainter lines may also be known, but are more likely to be unidentified at the start of

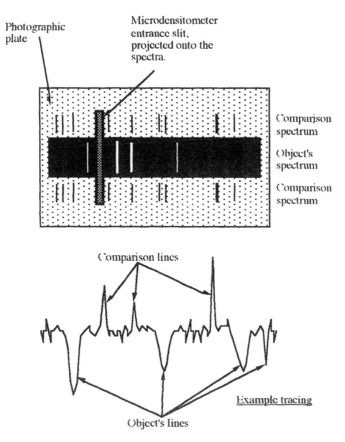

Figure 12.2   Using a long slit on a microdensitometer to obtain the comparison lines and the object's lines on a single tracing (spectrum shown as a photographic negative).

work on a spectrum.   The observer is therefore faced with the problem of determining which elements, and in what levels of ionization and excitation, are producing these lines. Line identification can never be wholly certain, and there may remain unidentified lines, or lines with ambiguous identifications, no matter how much effort is put into the identification process. The following procedure, however, is likely to lead to reasonable success in identification.

1. Decide which features in the spectrum are lines and which noise. This is best done by comparing two or more spectra to find which features repeat (lines) and which do not (noise).

2. Measure the wavelengths of all features estimated to be lines (see earlier).

3. From the literature or any other sources, identify as many lines as possible.

4. Determine the radial velocity of the object, and correct all the lines' wavelengths to their rest values.

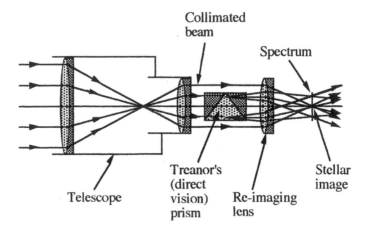

Figure 12.3 Treanor's approach to radial velocity determination.

5. From a set of tables of multiplets and their wavelengths (appendix C) find the atoms and ions from which lines have already been identified in the spectrum. List the other lines over the spectral region covered by the spectrum that come from those same atoms and ions. Check these lines against the wavelengths of the unidentified lines in the spectrum. If any of the values agree to within the uncertainty of the wavelength determinations, then there is a good chance that the unknown line has been correctly identified. Especial attention should be paid to resonance lines (lines originating from the ground state) and lines from low multiplets since these are likely to be relatively strong. Relative intensities within multiplets (chapter 4) can also be used as a cross-check. Thus if an identification has suggested one of the weaker lines in a multiplet, but the stronger lines from that multiplet are not found, then the original identification is probably wrong. On the other hand if several or all the lines of a multiplet are found and in about their correct relative intensities (remember the rules giving intensities within multiplets are very approximate), then this is as close to a certain identification of those lines as you are likely to get.

6. Repeat step 5 for the same atoms and ions but for one stage of ionization either side of those so far identified in the spectrum.

7. Return to step 4, if by now a significantly more precise value for the radial velocity may be found.

8. Check any remaining unidentified lines against a catalogue of line wavelengths (appendix C), and list all the lines to be found within the measurement uncertainty limits for the observed lines. This can sometimes be a large number of lines.

9. Eliminate improbable lines from the lists. This is a matter of judgement, and may well be wrong the first time round. It requires assessment of many items of information. Thus an estimate of the likely temperature of the object

should be obtainable from those lines already identified. Molecular bands are then unlikely candidates for an object which contains lines from doubly or triply ionized atoms, or lines due to C IV in a spectrum known to contain lines of neutral metals etc. The normal abundances of elements can also be used. Thus, although a few stars do have strong lines of europium or praseodymium etc, the observer is unlikely to make many errors in rejecting them as initial candidates for identifications. Forbidden lines should not be expected unless dealing with interstellar nebulae. Lines from high multiplet numbers or intercombination lines are on the whole less likely than lines from lower multiplet numbers or from normal transitions, and so on.

10. Look for any repetitions amongst the remaining lines in the lists (i.e. two or more lines from the same element, same ion, same multiplet etc). With these potential identifications return to step 5. If there are no repetitions, or when they have been exhausted, return individually to step 5 for each potential identification for each unknown line. At this stage, or earlier, it may have become clear that some features initially judged as noise should now be taken as lines, or vice versa, and a return to step 1 is indicated.

11. If, after several iterations, some lines remain unidentified, then return to step 9 and consider progressively less probable identifications. If the object is an interstellar gas cloud, or could include rarefied gases in an outer shell, then forbidden lines from metastable levels will also need to be considered.

12. Finally, consider that some lines may be shifted by a different radial velocity from that of the object; for example, lines from the interstellar medium or from a companion in a binary system. In such cases the lines are likely to be representative of quite different levels of excitation and ionization from the lines from the object itself. There are also symbiotic stars wherein lines from apparently different conditions do coexist.

13. Unfortunately when all the above has failed, you can no longer invent a new element! Though this was an option open to Norman Lockyer who discovered helium in this manner in 1868. Any remaining unidentified lines nowadays must remain so.

13

Spectrophotometry

13.1 INTRODUCTION

Spectrophotometry is the detailed study of a spectrum, looking at precise values for the line strengths and at the shapes of line profiles. Line profiles in particular are affected by innumerable processes: pressure, thermal motions, magnetic fields, rotation, expansion/contraction, turbulent and convective motions, limb darkening, gravity darkening, binarity, etc and can therefore, in principle, return data on all those effects. Potentially then, spectrophotometry can provide us with a wealth of information about astronomical objects. In practice, it may not be possible to separate out the various effects, and noise will always impose a greater or lesser degree of uncertainty upon the interpretations.

13.2 SPECTRAL CALIBRATION

As the name of the technique implies, spectrophotometry involves the measurement of absolute or relative intensities at different wavelengths. However, the response of all detectors varies with wavelength, so that comparing intensities across a spectrum requires this effect to be corrected. With photographic emulsions, the spectral response will normally be determined as a part of the photometric calibration (see below). With CCDs and other modern detectors, the spectral response will normally be determined by the manufacturer and supplied as a part of the data accompanying the instrument. The Earth's atmosphere plus optical components such as lenses, mirrors, filters and their coatings will also have a wavelength-dependent transmission or reflection, and these will need to be taken into account to find the overall spectral response of the detector–spectroscope–telescope combination. If necessary, the response can be determined experimentally by observing a source of known spectral distribution. This could be a well-known astronomical source, or an artificial source. In the latter case, an artificial black body is probably the best choice since its spectrum can be calculated precisely, provided that its temperature is known.

13.3   PHOTOMETRIC CALIBRATION

With many modern detectors, such as CCDs, the output is linear and directly proportional to the intensity between the noise level and saturation, or with the IPCS, directly proportional to the number of photons. Such detectors do not therefore need a photometric calibration unless absolute measurements are required, when they will have to be calibrated against a standard source. Photomultipliers, when used as photon counters, are also linear in their response. But at higher fluxes, when they act as analogue detectors, they can become nonlinear. The nonlinearity, however, can be reduced to $x\%$ or less by having the signal current no more than $x\%$ of the string current (the current to the dynode chain). If x is made small enough, the photomultiplier is then also effectively linear in its response.

The photographic plate by contrast is highly nonlinear in its response. Since photography is likely to continue in use for some spectroscopic purposes for some time to come, we need to look at its response and how it may be corrected. A graph of the response of the emulsion to different intensities is called the characteristic curve of the emulsion (figure 9.9). It may be obtained by photographing a standard source which contains a number of different intensities and whose relative intensities are known. Usually that standard source is in the form of a low dispersion spectrum (figure 9.8), so that the spectral response of the emulsion (above) may also be determined. The characteristic curve of an emulsion is affected by many things. A photometric comparison has therefore to be obtained for almost every spectrum plate. In particular the photometric comparison must be obtained on emulsion from the same batch and stored under the same circumstances as those used to obtain the spectrum, and both plates must be processed together under identical conditions. Other desirable precautions would include adjusting the intensity of the comparison source so that the exposure for the comparison is comparable in length with that for the spectrum, and providing an interrupted exposure for the comparison if that is the nature of the spectrum exposure through the use of a rocker (figure 9.2). If several similar spectra are obtained in quick succession on emulsion from a single batch, then a single photometric comparison will suffice for them all, provided that all the plates are processed simultaneously.

Once the characteristic curve for the emulsion has been obtained, then it is a simple matter in principle to convert the observed photographic densities in the image back to relative intensities. Except over the shortest sections of a spectrum, however, such a correction would be highly tedious to undertake by hand. Normally, therefore, the output from the microdensitometer that is tracing the spectrum is fed directly to a small computer. The characteristic curve is also fed into the computer, and the corrected spectrum output from the computer to a chart recorder or stored for further processing.

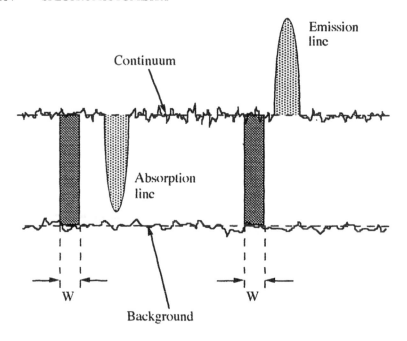

Figure 13.1 Equivalent width.

13.4 LINE STRENGTHS

Much of spectrophotometry is concerned with the detailed study of the shapes of line profiles. But for some purposes just the intensities of the lines are needed. These are conventionally measured as the Equivalent Width, W. This is the width of a nearby section of the continuum which has the same area as the line (figure 13.1). It has the advantage of being independent of exposure and spectroscope properties because it is normalized to the continuum. It is often also normalized by dividing by the wavelength to give W_λ, which may then be used to compare widely separated lines. On a good photographic spectrum, after photometric correction, the uncertainties in the equivalent width may be 5% to 10% . On a spectrum from a CCD or IPCS, 1% accuracy for W may be obtained on good spectra. Sometimes, however, the uncertainties may be much larger than these figures, such as for blended lines, when the noise level in the spectrum is high, or the number and/or width of the lines makes the position of the continuum uncertain.

13.5 LINE BROADENING

As already mentioned, many factors influence the shapes and widths of spectral lines, and these come under the general heading of line broadening. The intrinsic or natural shape of the lines has already been discussed (equations (4.21) and (4.24), figure 4.7); it is in the form of a Lorentz or damping profile and given by an equation of the form

$$I(\Delta\lambda) = \frac{A}{\Delta\lambda^2 + B} \qquad (13.1)$$

where $\Delta\lambda$ is the distance in wavelength terms from the line centre and A and B are constants. The whole half-width of the line (figure 4.8) is given by $(2B^{1/2})$.

For most purposes in astronomy the natural profile is too narrow to be observed, except for interstellar Lyman-α. Other factors therefore normally dominate the lineshape. Here we briefly review some of the more important of those factors, and look at the lineshapes that they produce.

13.6 PRESSURE BROADENING

Pressure broadening has already been mentioned in connection with the luminosity classification of stars (chapter 11) and is often the dominant broadening mechanism in stellar spectra. It arises because an atom, ion or molecule will be influenced by the nearby presence of other atoms, ions, molecules, free electrons or free nuclei. In the limit of a solid, this results in the formation of the valence and conduction bands etc (chapter 7). For a gas sufficiently rarefied to be producing line or band spectra, the effect of the influence of other particles is to perturb (change) the energies of the levels (see Zeeman and Stark effects etc, chapter 6). This alone would result in a broadening of an actual spectral line, produced by the effects of many atoms (or ions or molecules; from henceforth in this chapter, we shall use the term 'atom' normally to include ions, and much of the time molecules as well), since each atom would have its energy levels disturbed to a different degree. Additionally, however, at stellar surface temperatures, the velocities of the free electrons are sufficiently high that they will move large distances during the time it takes an atom to absorb or emit a photon (about 10^{-8} s). The effects of a free electron on the energy levels will thus change, and several such encounters may occur, during the absorption or emission process. Since, at constant temperature, the distances between particles reduce as the pressure increases, the effects of other particles on the energy levels will increase with pressure, resulting in the name given to this broadening mechanism.

By assuming that free electrons have an instantaneous effect on the energy levels during a close encounter with an absorbing or emitting atom, and that the heavier particles produce an electric field that is constant during a transition but varies from one transition to another, it is possible to calculate the line profile resulting from pressure broadening in the case of hydrogen and hydrogen-like ions. For other atoms, accurate calculations of their pressure-broadened line profiles are not possible. Various approximations, however, may be used,

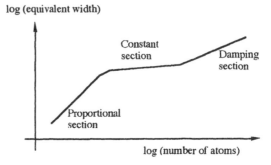

Figure 13.2   A schematic curve of growth.

and of course the lines may be observed in the laboratory. For all atoms therefore it is found that the pressure-broadened line profile is a Lorentz profile (equation (13.1)), though invariably much broader than the natural profile.

The pressure-broadened line only has a Lorentz profile shape if it is an emission line or a weak absorption line. Clearly the intensity at the centre of an absorption line cannot become less than zero. The stronger lines therefore saturate at their centres and their profiles become modified. This modification will also apply to the other types of line profile described below. The way in which a line increases in strength as the number of atoms producing it increases is described by the curve of growth (figure 13.2). This is the basis of the coarse analysis approach to determining element abundances, and will be encountered again in chapter 14. The curve of growth has three sections, representing different ways in which the line strength varies with the number of atoms. Initially, when the line is weak, its strength increases in direct proportion to the number of atoms (the proportional section). As the centre begins to saturate, the line strength becomes almost constant even though the number of atoms continues to increase (the horizontal section). Finally, when the density becomes high enough, pressure broadening widens the line and it starts to develop very broad wings and to increase in strength again. This section is known as the damping section, because the line has a damping (Lorentz) profile, modified by its central saturation. Over this section the line strength increases as the square root of the increase in the number of atoms producing it. These various effects on the line are shown schematically in figure 13.3.

13.7   DOPPLER BROADENING

After pressure, the broadening of lines due to Doppler shifts is the most commonly occurring effect. Such broadening occurs when atoms contributing to a spectral line have different line-of-sight velocities, so that their individual contributions are shifted by differing amounts from the normal wavelength. In all astronomical objects, thermal motions, turbulence and other fine scale

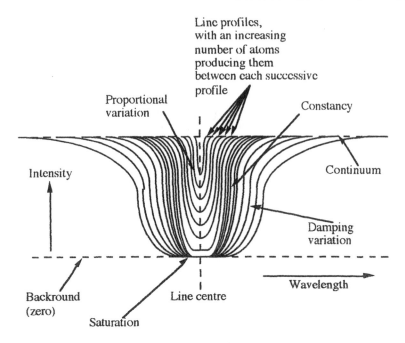

Figure 13.3 Change in a line profile with the increase in the number of atoms producing it.

processes can broaden lines. In stars and other angularly unresolved objects, macroscopic effects such as convection, rotation, expansion and contraction can also change the profile.

13.7.1   Thermal broadening

Thermal broadening arises from the components of the thermal motions of the atoms along the line of sight. In non-degenerate matter, the distribution of such velocities is given by the Maxwell–Boltzmann distribution:

$$\frac{N(p)\,\mathrm{d}p}{N} = \left(\frac{2}{\pi m^3 k^3}\right)^{1/2} \frac{p^2}{T^{3/2}} \mathrm{e}^{-p^2/2mkT}\,\mathrm{d}p \qquad (13.2)$$

where $N(p)$ is the number density of particles with momenta in the range p to $p + \mathrm{d}p$, N is the total number density of the particles and m is the mass of the particle. The resulting line profile for emission lines and faint absorption lines is thus given by

$$I(\lambda) = I(\lambda_0)\mathrm{e}^{-mc^2(\lambda_0-\lambda)^2/2kT\lambda^2} \qquad (13.3)$$

where λ_0 is the rest (zero velocity) wavelength (note different usage from chapter 12) and λ is the wavelength at the point within the line being considered, or since the linewidths are small compared with the wavelength,

$$I(\lambda) \approx I(\lambda_0)e^{-mc^2(\lambda_0-\lambda)^2/2kT\lambda_0^2}. \tag{13.4}$$

This profile is of Gaussian shape, and has a half-half-width of

$$\Delta\lambda_{1/2} = \left(\frac{2kT \log_e 2}{mc^2}\right)^{1/2}\lambda_0. \tag{13.5}$$

For medium mass atoms the thermal broadening in the optical region is thus about 0.01 nm at a temperature of 10 000 K, about a thousand times greater than the natural linewidth.

13.7.2   Turbulent broadening

Turbulence, including convective motions, produces a Gaussian distribution of velocities for the atoms similar to that for the thermal motions, but with an average turbulent velocity, V, replacing the rms thermal velocity, $(3kT/m)^{1/2}$. The turbulent line profile is thus given by

$$I(\lambda) \approx I(\lambda_0)e^{-c^2(\lambda_0-\lambda)^2/V^2\lambda_0^2} \tag{13.6}$$

and its half-half-width by

$$\Delta\lambda_{1/2} = \left(\frac{V^2 \log_e 2}{c^2}\right)^{1/2}\lambda_0 \tag{13.7}$$

so that an optical line would have a width of about 0.005 nm for turbulent and/or convective velocities averaging 2 km s^{-1}.

13.7.3   Combined profiles

Pressure, thermal and turbulent broadening effects are often found in combination. The combined effect of the two Gaussian profiles produced by thermal and turbulent motions is just a third Gaussian. The contributions from the two effects therefore cannot be distinguished on the basis of the line profile alone. The combined profile is given by

$$I(\lambda) \approx I(\lambda_0)e^{-c^2(\lambda_0-\lambda)^2/(2kT/m+V^2)\lambda_0^2} \tag{13.8}$$

and its half-half-width by

$$\Delta\lambda_{1/2} = \left(\frac{(2kT/m + V^2) \log_e 2}{c^2}\right)^{1/2}\lambda_0. \tag{13.9}$$

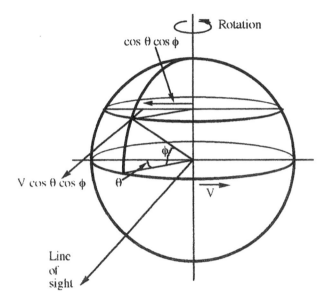

Figure 13.4   A rotating star.

The combination of Lorentz and Gaussian profiles results in the Voigt profile. Tables of such profiles may be found in *Astrophysical Quantities* by C W Allen (appendix C). That reference book also gives tables to enable the individual Lorentz and Gaussian components to be separated out from an observed Voigt profile. The effects of pressure broadening may thus be found separately from those of the thermal and turbulent motions.

13.7.4   Rotational broadening

Many of the hotter stars rotate rapidly. Since their discs are unresolved, we see contributions to a spectral line from both the approaching and receding parts of the star simultaneously. For a uniformly bright star, with thin surface layers, and ignoring other broadening mechanisms, the profile for pure rotation is readily obtained. The line-of-sight velocity of any point on the surface, when the rotational axis is perpendicular to the line of sight, is easily seen to be (figure 13.4)

$$V(\theta, \phi) = V \cos \theta \cos \phi \qquad (13.10)$$

where the coordinate system is shown in figure 13.4.

From the diagram, the projected distance from the rotational axis of that same point is also easily seen to be $\cos \theta \cos \phi$. The line-of-sight velocity, and therefore the Doppler shift, for any point on the surface is thus just proportional to the projected distance of that point from the rotational axis. The line profile is thus given by

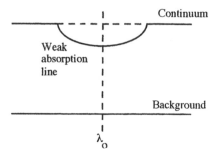

Figure 13.5 Line profile for pure rotation.

$$I(\lambda) \approx I(\lambda_0) \left(1 - \frac{c^2(\lambda_0 - \lambda)^2}{V^2\lambda_0^2} \right)^{1/2} \tag{13.11}$$

and the half-half-width by

$$\Delta\lambda_{1/2} = \frac{\sqrt{3}V}{2c}\lambda_0. \tag{13.12}$$

For an optical line, and a projected equatorial velocity of $200\,\mathrm{km\,s^{-1}}$, the half-half-width is thus about 0.3 nm. The line profile shape is semi-elliptical (figure 13.5). It is not too difficult to show that a similar line shape results whatever the inclination of the rotational axis. A semi-elliptical line profile thus indicates that the principal broadening mechanism is rotation. The projected equatorial velocity, $v \sin i$, is then obtained directly from the whole width of the line, since the extreme edges of the lines correspond to light originating at the limbs of the star.

13.7.5 Expansion and contraction

Some stars, such as novae and supernovae during their explosive phase(s), have monotonically expanding surfaces. Others, such as the Cepheids, may expand and contract. Much more rarely, some protostars may be observed to have monotonically contracting surfaces. The YY Orionis stars are a possible example of this latter case. Planetary nebulae and H II regions are also expanding, but may usually be angularly resolved, so that the lines in individual spectra will just exhibit wavelength changes due to the Doppler shifts appropriate to those parts of the nebula that are being observed, rather than changes to the profiles.

A different physical situation which results in similar line profiles to those produced by expansion is that of mass loss. Many stars are losing mass in the form of stellar winds. In the case of the hotter stars, the amount of material in the wind may be considerable, and the spectrum may be dominated by lines from the wind material rather than from the stellar surface. Emission lines formed in the material of the wind will have expansion-type line profiles, though in many cases modified by acceleration or deceleration of the material over the emission region. Absorption lines produced by the wind will again have similar line profiles, but with only those parts of the profile produced by the material silhouetted against the stellar photosphere represented. In practice, this means that absorption lines from stellar winds will have fairly normal profile shapes, but shifted to shorter wavelengths. Often many factors influence the line profiles resulting from winds and they must be modelled on an individual basis.

Objects ranging from protostars to active galactic nuclei may be losing matter in the form of jets. For angularly unresolved objects, there will clearly be an effect on the shape of those line profiles originating from material in the jets. The situation, however, is too complex to give any general rules, and such objects will normally have to be modelled individually, to take account of factors such as the opening angle of the jets, their velocity distribution, their angle to the line of sight and the possible obscuration by other material of the rearward jet.

In the simple case of spherically symmetric expansion or contraction, however, we may obtain an analytical solution to predict the resulting line shape, and this will provide a useful guide to what may be expected in more complex situations. For a shallow (i.e. negligibly thick compared with its size) absorbing or emitting region such as may be the case with Cepheids, the physical situation is the same as that shown in figure 13.4, except that the velocity is radial to the star at each point on its surface. The velocity along the line of sight at a particular point is thus given by $V \cos \phi \sin \theta$, where V is the velocity of expansion or contraction. The resulting line profile for weak absorption lines or emission lines without self-absorption is triangular (figure 13.6(a), and given up to the maximum wavelength shift ($\Delta\lambda = \lambda_0 V/c$) by

$$I(\lambda) = I(\lambda_0) \left(1 - C\frac{\lambda_0 - \lambda}{\lambda_0}\right) \qquad (13.13)$$

where C is a constant. The half-half-width is thus

$$\Delta\lambda_{1/2} = \frac{\lambda_0}{2C}. \qquad (13.14)$$

For stellar winds and many other types of expansion, the absorbing region has a significant thickness in comparison with its size. The absorption at a particular wavelength within the line profile is then the result of contributions from a broad annulus of the disc of the star centred on the line of sight (figure 13.7). The lineshape remains roughly triangular, but with the sides now segments of parabolas (figure 13.6(b)). The line profile is given by

$$I(\lambda) = I(\lambda_0) \left[1 - C\left(\frac{\lambda_0 - \lambda}{\lambda_0}\right)^2\right] \qquad (13.15)$$

for

$$\lambda > \lambda_0 \left(1 - \frac{V}{c}\sqrt{1 - \frac{r_0^2}{r_1^2}}\right) \qquad (13.16)$$

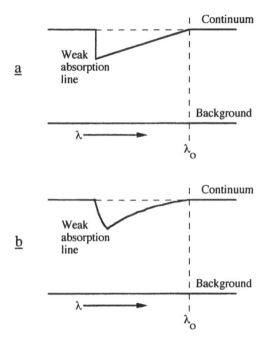

Figure 13.6 Weak absorption line profiles for (*a*) a thin absorbing region and (*b*) a thick absorbing region.

and

$$I(\lambda) = I(\lambda_0) \left\{ 1 - \left(\frac{r_1^2}{r_0^2} - 1 \right) \left[1 - C \left(\frac{\lambda_0 - \lambda}{\lambda_0} \right)^2 \right] \right\}$$
(13.17)

for

$$\lambda \leqslant \lambda_0 \left(1 - \frac{V}{c} \sqrt{1 - \frac{r_0^2}{r_1^2}} \right)$$
(13.18)

where r_0 and r_1 are the inner and outer radii of the absorbing region (figure 13.7).

Real situations are likely to be much more complex than these simple models and require the inclusion of limb darkening and other broadening mechanisms, but asymmetry in an observed line profile may usually be regarded as a sign that expansion or contraction is occurring.

An emitting region that is large compared with the star leads to rather different line profiles from those for the absorption lines, because if it is optically thin, then contributions to the line are received from both the approaching and receding sections. The line is therefore symmetrical around the unshifted wavelength. The profile is rectangular in the case of optically thin regions (figure 13.8(*a*)). If absorption as well as emission is important throughout the region, then the emission profile becomes parabolic (figure 13.8(*b*)).

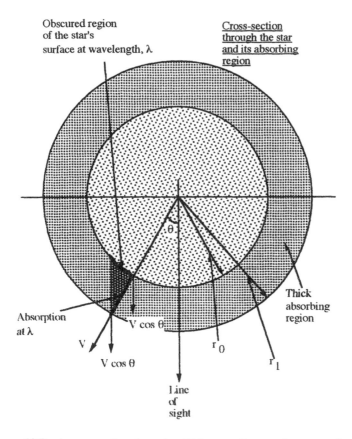

Figure 13.7   A cross section through a thick expanding envelope around a star.

In regions that are accelerating as they expand outwards, absorption may be produced by the outer extremities of the region, while emission comes from the inner parts. The line profiles shown in figures 13.6 and 13.8 may then become combined. The absorption occurs at the short wave edge of the emission, and results in a line shape known as the P Cygni profile after one of the brighter stars with this type of spectral line (figure 13.9). Very occasionally the inverse effect may occur for infalling material and produce reversed P Cygni lines.

13.7.6   Binarity

Close binary stars often exhibit evidence of their nature through the splitting of their spectral lines, and from the to and fro motion of each line component about its mean position in the spectrum over the orbital period. In a low dispersion spectrum, such lines may not be seen separately, but as a single, broader line whose width varies periodically. Some binaries, such as the W UMa stars, have

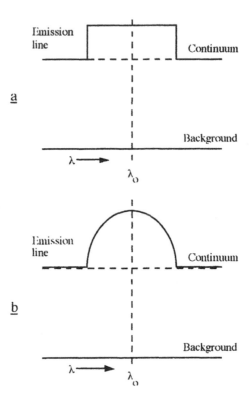

Figure 13.8 Emission line profiles from a thick expanding region. (*a*) Optically thin case; (*b*) absorption as well as emission occurring throughout the region.

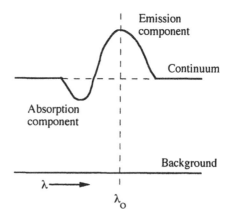

Figure 13.9 P Cygni line profile.

such small physical separations that the components touch each other, and they may even possess a common photosphere. In the latter case we have effectively a single star with two cores, and even the highest resolution spectroscopes will show only single lines with variable width. The precise interpretation of the line profiles will require individual modelling, though this may be simplified because the component stars in close binaries are often very similar to each other, being the result of the fission of a larger object during the last stages of its formation. The observed lines will therefore be formed out of identical components from each of the stars.

13.8   MAGNETIC FIELDS

We have seen (chapter 6) that emission or absorption in the presence of a magnetic field results in lines being split (the Zeeman effect). Although the magnetic fields in some stars can be quite strong, the stars are not angularly resolved and therefore even with a well ordered field we receive light from a wide range of angles of that field to the line of sight. In many cases, the field will not be well-ordered but will be complex like the general magnetic field of the Sun. The overall effect of Zeeman splitting on the observed lines is thus not to split but to broaden the lines because they comprise many different Zeeman components with many different individual separations and intensities.

In addition to splitting lines, the Zeeman effect results in the different components being polarized in different ways. In the wings of a line affected by a strong magnetic field, some of this polarization will remain, and it can be used to estimate the magnetic field strengths, down to about values of 0.01 T.

13.9   OTHER EFFECTS

13.9.1   Limb and gravity darkening

The brightness of the visible disc of the Sun decreases from its centre to its edge because we see to different depths in the solar atmosphere. Similar effects occur in most stars, and in the giants to a much greater degree than in the Sun. A separate effect, not noticeable for the Sun, is that of gravity darkening. The brightness of a point on the surface of a star is directly proportional to the local gravitational acceleration (von Zeipel's law). In close binaries and rapidly rotating stars, the effective gravity will vary over the surface, and so, by this theorem, will the surface brightness. Neither of these effects will result in line broadening; indeed they may cause the lines to be narrower than would otherwise be the case. They do, however, have to be taken into account when interpreting line profiles.

13.9.2   Surface features

Features analogous to sunspots may be found on many other stars, and other types of surface changes, such as the variation in the abundances of some elements over the surfaces of Ap stars, may occur. These will clearly affect the detail of line profile shapes. When the star is rotating reasonably rapidly, the modifications of a spectral line due to individual features may be tracked as they move through the line throughout a rotational period. They may then be individually determinable. Maps of the surfaces of stars may be produced in favourable cases by this technique, which is known as Doppler imaging.

13.9.3   Isotopes and other 'atomic' effects

In very narrow lines, such as those from the interstellar medium, splitting of the line because of the slightly different energies of lines from different isotopes of the same element may be observable. This is particularly the case for the lighter elements. Similarly, hyperfine splitting (chapter 3) and rotational splitting (for molecular lines, see chapter 5) may sometimes be detected. When these effects are unresolved or the components blurred into each other by other broadening effects, then they add to the overall broadening of the lines.

13.9.4   Instrument

The resolution of spectroscopes is limited by many factors (chapter 8). The core of the instrumental profile represents the narrowest observable spectral line. Lines narrower than this will be seen with a profile due to the effects of the spectroscope, and not due to any properties of the object being observed. Some reduction in the effect of the instrument may be achieved by deconvolving the instrumental profile. The improvement effected by this process, however, is fairly limited and it is left for the reader to investigate in more specialized books than this one (appendix C).

14

Stars

14.1 INTRODUCTION

This chapter is not concerned primarily with stars themselves, but with the way spectroscopy can be used to obtain information about the stars. Several aspects of this have already been covered: spectral types (chapter 11), radial velocities (chapter 12), and pressure, turbulence, rotation, expansion etc (chapter 13). Here, therefore, we are concerned with some of the remaining aspects of stellar spectroscopy. The topics covered are those which find widespread application, but the list is far from complete. There are many ingenious and subtle uses of spectroscopy, often for special and one-off cases, to be found in the literature, and for which there is insufficient room here.

14.2 DISTANCE

The most fundamental way of determining the distances of stars is by observing the parallax shifts due to the Earth's displacement in space as it moves around its orbit, and therefore does not involve spectroscopy at all. However, parallax can only be used to obtain reasonably reliable distances out to about 50 pc (shortly due to be extended to a few hundred parsecs when the results from the Hipparcos spacecraft become available). Greater distances than this are determined by a variety of methods which form a hierarchy with parallax as the base. The first tier then includes those methods of distance measurement which may be calibrated using the parallax determinations. The second tier is calibrated from the results from the first tier, and so on. Several of these later approaches are based upon spectroscopic measurements.

14.2.1 Spectroscopic parallax

This is a somewhat inappropriately named technique, for what is actually one of the 'Standard Candle' methods of finding the distances of stars, and therefore finds the distance directly without determining the parallax itself. Standard candle methods of distance determination are based upon being able to measure or estimate the absolute magnitude of the star, measure its apparent magnitude, and then find the distance using the relation

$$D = 10^{0.2(m-M+5)} \quad \text{pc} \tag{14.1}$$

where M is the absolute magnitude and m is the apparent magnitude. In spectroscopic parallax, the absolute magnitude of the star is found from its position on the H/R diagram (figure 14.1), using accurately determined spectral and luminosity classes (chapter 11). The result for an individual star may contain large uncertainties, though for many stars it may be the only way of estimating the distance.. However, when the method is applied to a cluster, the member stars of which are all effectively at the same distance from us, it can become reasonably accurate.

14.2.2 Wilson–Bappu method

This is another of the standard candle methods, and uses the empirical relationship that is found between the widths of the emission cores of the Ca II H and K lines and the star's absolute magnitude. The relationship is

$$M_V = 27.6 - 14.9 \log_{10} w \tag{14.2}$$

where w is full width of the emission core at half its maximum, measured in km s^{-1}.

14.2.3 Expanding nebulae

Novae, supernovae, the central stars of planetary nebulae, and the stars embedded in H II regions are all physically associated with gaseous nebulae that are expanding. If the expansion is isotropic, or may be assumed to be so, then the distance of the nebula, and hence that of the star, is obtainable by taking the velocity of expansion along the line of sight, determined spectroscopically, to be the same as that across the line of sight, determined by direct observations (chapter 16).

14.2.4 Moving cluster method

The stars of a galactic cluster are gravitationally bound together and moving through space with similar paths. Seen from the Earth, perspective causes the velocities of the stars in the cluster to seem to diverge from or converge to a single point on the celestial sphere. The radial velocity, v_r, proper motion, μ, and angular distance from the radiant or convergent point, β, of a star in the cluster are then related to its distance by

$$D = \frac{v_r}{5\mu} \tan \beta \quad \text{pc} \tag{14.3}$$

Absolute magnitude

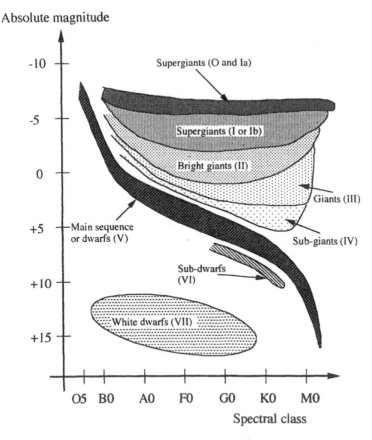

Figure 14.1   The Hertzsprung–Russell (H/R) diagram.

where v_r is measured in $km\,s^{-1}$ and μ in arcsec/yr.

14.2.5   Galactic rotation

The rotation of the Galaxy leads to the radial velocity of a star depending upon its distance. The intrinsic motions of stars make this a very imprecise method if applied to a single star, but for star clusters it can be a little better. The mean distance of a particular type of star can also be estimated by this approach.

14.3   TEMPERATURE

The surface temperature of a star and its spectral type are very closely related (figure 11.1), so that an accurate determination of the spectral type will give a value for the effective temperature. Temperature also appears in the Boltzmann and Saha equations (equations (4.5) and (11.1)), so that the levels of excitation and ionization determined from the lines observed to appear in the spectrum can lead to estimates of the excitation and ionization temperatures. Similarly the kinetic temperature may be determinable from the thermal Doppler broadening of the spectral lines (equations (13.4) and (13.5)).

For many stars, the continuous part of the spectrum closely approximates the shape of a black body spectrum, especially at the longer optical wavelengths. The temperature of the star may therefore be found by fitting the continuum to a black-body curve, given by the Planck equation:

$$B_\lambda(T) = \frac{2hc^2\mu^2}{\lambda^5 \left(\exp(hc/\lambda kT) - 1\right)} \tag{14.4}$$

where $B_\lambda(T)$ is the intensity of the radiation at wavelength λ for unit wavelength interval and μ is the refractive index of the medium (usually very close to unity).

Care needs to be taken with this approach, however, since the envelope of the observed spectrum can become seriously distorted at shorter wavelengths by ionization edges, and over all parts of the spectrum if there are numerous spectral lines (line blanketing). The latter problem is particularly severe for the cooler stars with their numerous molecular bands. For any star the calibration of intensities over a large wavelength interval (chapter 13) must be done accurately if the method is to work. The determination of temperature up to about $10\,000$ K from the B–V colour index via the formula

$$T \approx \frac{8540}{(B - V) + 0.865} \quad K \tag{14.5}$$

is another but non-spectroscopic example of this method.

Finally, the effective temperature of a star can be determined by equating its total luminosity to that of a similarly sized black body if intensity determinations over a wide spectral range are available. From the Stefan–Boltzmann equation, we have

$$T_{\text{eff}} = 34.4L^{0.25}R^{-0.5} \quad K \tag{14.6}$$

where L is the star's bolometric luminosity in watts and R is the star's radius in metres.

14.4   ELEMENT ABUNDANCES

Clearly, all other things being equal, the more atoms of an element that are present in a star's atmosphere, the stronger will be its spectral lines. The variation of line strength with the number of contributing atoms is called the curve of growth (figure 13.2). Using the curve of growth to determine abundances is a process known as coarse analysis.

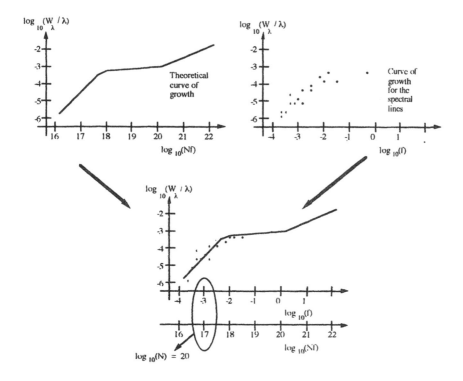

Figure 14.2   Schematic element abundance determination from the curve of growth.

In order to determine the abundance of an element, a generalized curve of growth is needed which applies to many or all of the atom's lines. This may be accomplished in several ways, but probably the most widespread approach is to normalize the equivalent width and the number of atoms. The equivalent width is normalized by dividing it by the wavelength of the line. Thus $\log_{10}(W_\lambda/\lambda)$ is plotted as the ordinate of the curve of growth. The abscissa is then $\log_{10}(Nf)$, where f is the oscillator strength (chapter 4) and Nf is thus the effective number of atoms contributing to the formation of the line.

The observations are plotted onto a graph of $\log_{10}(W_\lambda/\lambda)$ against $\log_{10}(f)$. Since the oscillator strengths can vary by large factors from one line to another, the resulting curve will also be a curve of growth. The shift along the abscissa required to superimpose the theoretical and observed curves of growth then gives the value of $\log_{10}(N)$ directly (figure 14.2).

The relationship of N to the abundance of the atom depends upon the lines used in the determination. If they originate from an excited level then N will be the number of atoms excited to that level etc. If TE or LTE can be assumed then the Boltzmann equation (equation (4.5)) can be used to determine the populations of the other levels, and so the total number of atoms or ions. Saha's equation

(equation (11.1)) can then be used to relate this to the numbers of atoms in other states of ionization, and so the total abundance of the element found. Alternatively, the populations of the other levels and ionization stages may be determinable directly from the lines produced by them.

A more sophisticated approach to abundance determination is known as fine analysis. Fine analysis requires much higher quality data, and the corrections for instrumental degradation etc must be carefully applied. The technique uses a detailed model atmosphere, together with assumed abundances, temperature, pressure, turbulence and surface gravity to predict line strengths and line profiles. These are then compared with the observations, and the assumed parameters altered until a good fit is achieved. The initial values would normally be obtained from a coarse analysis. Once fine analysis has been used on one star, coarse analysis is often adequate to give high quality relative compositions in another star of similar type.

Fine analysis, as it becomes more detailed, gradually merges into spectral synthesis. The spectrum is then calculated point by point over a wide range of wavelengths for comparison with the observations. This is a very complex procedure since the usual simplifying assumptions (LTE etc) have to be abandoned. The number of reliable models of stars based upon spectral synthesis is relatively few at the moment, but this is undoubtedly the direction in which spectroscopic abundance analysis will proceed in the future.

14.5   VARIABLE STARS

Almost any property of a star may be found to change, leading to visual binaries, photometric variables, spectroscopic variables, etc. The variation of one parameter frequently leads to other changes so that variable stars may belong to several of these sub-classifications. Changes in the spectrum can arise from many factors, and can therefore conversely provide information on those factors. The more commonly encountered types of spectroscopic variable are reviewed below.

14.5.1   Spectroscopic binary stars

Unless the plane of the orbit is close to perpendicular to the line of sight, the Doppler shift due to the orbital motions about their common centre of mass for the two stars in a binary system leads to changing wavelengths for the lines in the spectrum. When the two stars are within one or two stellar magnitudes of each other in brightness, the lines from each will be detectable in the combined spectrum. If they differ by more than about two magnitudes, then generally only one set of lines will be visible. If the two stars are of significantly different temperatures, however, lines from the hotter star may be visible in the visual region, while those from the cooler star become apparent in the infrared. When the same line is detectable from both stars, if the components of the orbital motion are high enough, it will be seen to split and recombine cyclically at half the orbital period of the binary (figure 14.3). When only one line is detectable, it will be seen to oscillate about its mean position (figure 14.3).

The observed velocities of the two stars are inversely proportional to their masses:

$$\frac{M_b}{M_a} = \frac{v_a(t)}{v_b(t)} \tag{14.7}$$

where M_a and M_b are the masses of the two stars and $v_a(t)$ and $v_b(t)$ are the observed components of their orbital velocities at time t.

Thus if the mass of one star may be estimated (perhaps from its spectral and luminosity class), then that of the other may also be found. If the orbital plane of the binary is close to the line of sight, so that eclipses occur, then the individual masses of the stars are determinable directly from Kepler's third law

$$M_a + M_b = \frac{4\pi^2 a^3}{G T^2} \tag{14.8}$$

where G is the gravitational constant $(6.670 \times 10^{-11}\,\mathrm{N\,m^2\,kg^{-2}})$, a is the semi-major axis of the orbit and T is the orbital period. This combination of spectroscopic and photometric observations of binary stars is the main way in which stellar masses are found.

14.5.2 Cepheids

Cepheids and related variables such as the RR Lyrae stars exhibit two types of spectral change which arise from their temperature changes and from their expansion and contraction. The size change leads to alterations in the shapes of the spectral lines (figure 13.6), with the asymmetry reversing between the expanding and contracting phases of the star's cycle. The temperature changes lead to changes in the observed spectral type of the star by as much as an entire spectral class (e.g. F5 to G5 and back for the brightest cepheids).

14.5.3 Cataclysmic variables

Novae and supernovae may exhibit small spectral changes outside the periods of their outbursts; dwarf novae for example are generally close binary systems. The main changes, however, occur during the outburst, when the spectrum changes from that characteristic of a star to that characteristic of an interstellar nebula (chapter 16). The change arises from the enormous expansion in the outer layers of the star that are thrown off during its explosion. The spectral type is initially that of a hot star, typically spectral class A or B. As the outburst progresses, the spectral type changes towards lower temperatures and emission lines start to appear. The emission lines then become stronger and stronger,

Double line
spectroscopic binary

Single line
spectroscopic binary

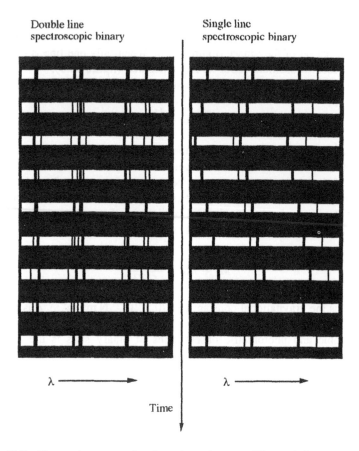

λ ⟶   λ ⟶

Time

Figure 14.3 Time series spectra showing schematic spectral line variations due to orbital motion in a binary star.

and the continuous component weakens, until the emission lines dominate the spectrum. For novae, the spectrum will eventually revert to that observed before the outburst as the ejected material expands to the point of transparency and the central stars are detected again. With supernovae, the core may become visible some time after the outburst as a pulsar at radio wavelengths, or very much more rarely, at optical and shorter wavelengths.

Flare stars also exhibit spectral changes during their outbursts, this time from their normal K and M spectral classes towards higher temperatures.

14.5.4 Rotational variations

Any rotating star with surface structure may exhibit changes as that structure rotates into and out of sight. Amongst the Ap stars, surface abundance variations lead to small changes in the spectral lines. This can be used to produce maps of the abundance variations over the star's surface (Doppler imaging, chapter 13).

14.5.5 Emission line variables

A small proportion of stars possess emission lines in their spectra. These include the Wolf–Rayet stars, Of stars, Be and Ae stars, T Tauri stars, young stellar objects and the cool supergiants. The emission lines are generally produced in an envelope of low density material surrounding the star. This material may have been ejected from the star or it may have been left behind during its formation. Frequently the material in the envelope will be inhomogeneous and may be expanding, contracting, jetting or orbiting around the star. Material may also be being added to the envelope or lost from it. The emission lines produced by the envelope material are therefore often variable in intensity and sometimes in their wavelength. Occasionally, the emission lines may change from single to double or vice versa. The details of these changes can reveal much about the nature of the envelopes around the stars, but are left for the reader to pursue in the specialist literature.

14.6 THE SUN

> CAUTION. Care should always be practised in any observations of the Sun. Its direct intensity is sufficient to cause blindness, or to damage the telescope or other instruments. Professionally produced solar filters should always be used on both the main instrument and any finding/guiding telescope, and their manufacturer's instructions carefully followed. Smoked glass and overexposed film are not suitable since they may transmit the infrared part of the spectrum.

The high flux of radiation from the Sun, plus the fact that it can be angularly resolved, have led to several specialized techniques being developed for its study, which are not applicable to other stars.

14.6.1 Spectrohelioscopes and narrow band filters

Observations of the Sun made at the centres of strong absorption lines will be of a higher level in the Sun's atmosphere than the white light photosphere. Such images of the Sun are generally known as spectroheliograms and can reveal much about the nature of the chromosphere and lower corona. They are obtained in two main ways. The simplest, conceptually, is just to interpose a very narrow band filter into the optical train of a normal telescope. Such filters have bandwidths of 0.05 nm or less and are usually centred on the H-α line, or the H or K lines of ionized calcium. Although the filters may have their central wavelengths slightly tunable through altering their temperature or by inclining them to the light beam, a separate filter will be needed for each spectral line, and they are expensive.

A more flexible system may be produced based upon a medium resolution spectroscope. This is known as a spectrohelioscope. It consists of a conventional spectroscope with the addition of a second slit superimposed onto the spectrum. That second slit may be moved until it isolates the desired absorption line. The light passing through it will then be an image in the light of the selected absorption line of that portion of the Sun which is falling onto the entrance slit of the spectroscope. The narrow-band image of the whole Sun may then be built up by trailing the white-light image over the entrance slit and recording the varying output from the second slit. More usually, however, the entrance slit is oscillated back and forth so that it covers the solar image rapidly. The second slit must then also be oscillated to maintain its registration onto the absorption line. If the oscillations are of a high enough frequency, then the narrow-band solar image may be observed directly using an eyepiece placed after the second slit. Alternatively an image may be obtained after refocusing the light beam using a CCD or by photography.

14.6.2   Magnetograms

Zeeman splitting (chapter 6) produces components to spectral lines which have differing polarizations. In Leighton's method two spectroheliograms are obtained, one in each of the wings of a Zeeman-broadened absorption line. A differential analyser, which consists of a quarter-wave plate and a suitably oriented calcite crystal, is used to isolate the stronger circularly polarized component of the radiation in each case. A composite print with one image as a negative and the other as a positive then produces a grey background where magnetic fields are weak and a light or dark feature where they are strong, providing a very vivid image of the pattern of the solar magnetic fields.

Another method is due to the Babcocks and uses a differential analyser with the quarter wave plate composed of ammonium dihydrogen phosphate. The birefringence of this material is induced by a high electric potential across it, and it may be reversed by reversing the potential. Because the predominant polarization varies across the line, there is a shift in the line's position as the wave plate is reversed. That shift is detected by a pair of photomultipliers accepting radiation in each of the wings of the line. Scanning the entrance aperture of the instrument over the solar disc provides a magnetogram of the whole Sun.

14.6.3   Surface features

Sunspots, flares, etc may be studied by relatively conventional spectroscopes. Because the Sun is an extended object, however, the outer wings of the instrumental profile (figure 8.11), which are weak but cover a large area, may seriously pollute the spectrum. A spectroscope with a very 'clean' instrumental profile, concentrating most of its energy into the central core, is therefore required for solar work.

14.6.4 Chromosphere

For a few seconds after totality begins during a solar eclipse and just before it ends, the solar photosphere is eclipsed, while the chromosphere remains visible. The spectrum of the chromosphere obtained in these moments is called the 'Flash' spectrum, because it is visible for such a brief period. The chromospheric spectrum consists almost entirely of emission lines, and it can therefore be studied with a slitless spectroscope. The flash spectrum is composed of a series of monochromatic images of the chromosphere in each of its emission lines (figure 1.4).

14.6.5 Corona

The solar corona is also best observed during a solar eclipse, but may additionally be studied using either Earth-based or space-based coronagraphs. Its optical spectrum may be observed with a conventional spectroscope and contains absorption lines due to scattered solar light, plus emission lines due to highly ionized atoms from the material of the corona itself.

15

Planets and other Minor Bodies of the Solar System

15.1 INTRODUCTION

Spectroscopy of the planets (and to avoid frequent repetition, the term planet will be taken in this chapter to include planetary satellites, asteroids, meteoroids and comets) can potentially reveal much information. However, spectroscopy has traditionally taken second place to direct imaging of the planets as the preferred mode of study, or in the case of the Earth, Moon, Venus and Mars, to direct sampling. Furthermore the strict application of the brief of *Optical* Spectroscopy for this book would limit the range of applications of spectroscopy that could be discussed in this chapter very considerably. We therefore extend the coverage to include parts of the ultraviolet, far infrared and even microwave regions where this seems particularly appropriate.

For the Earth and those planets with orbiting planetary probes, astronomy merges seamlessly with remote sensing. In order that this chapter does not become longer than the rest of the book, some selectivity has had to be applied in the choice of topics to be covered. Techniques and instruments applicable only to the Earth have therefore in general been omitted. This has resulted in the omission of such topics as Earth resources and meteorological spacecraft, multispectral scanners and other types of imaging radiometer, selective chopping radiometers, laboratory-based spectroscopy of samples, ionosonds etc, etc, and the interested reader is referred to other books for coverage of these topics (appendix C).

Except at microwave and longer wavelengths or for a few exceptional phenomena such as aurorae and lightning, the radiation that we receive from the planets is reflected or scattered solar radiation. The spectral features due to the planet are therefore superimposed upon the already complex solar spectrum, and the latter has to be subtracted from the observed spectra before the planetary effects may be studied.

We may conveniently divide the spectroscopic study of the planets into the studies of their surfaces and atmospheres. The spectroscopic techniques involved also depend upon whether they are Earth-based or spacecraft-based, and in the latter case upon whether the spacecraft is a fly-by, an orbiter or a lander.

15.2 PLANETARY ATMOSPHERES AND COMETS

Most spectroscopic studies of planetary atmospheres are concerned with molecular lines (chapter 5) and therefore involve the infrared and longer wavelengths. In a few special cases, however, visual atomic spectroscopy can still make a contribution. One example applying to an object usually thought to be without an atmosphere, and therefore showing the high sensitivity of spectroscopic techniques, may be found with our own Moon. Atomic sodium can be detected as a cloud around the Moon, out to several lunar radii and with densities of some few tens of millions of atoms per cubic metre, from the scattering of solar light in the sodium D lines at 589.0 and 589.6 nm. A similar cloud has been discovered around Jupiter by the same approach. The Swan bands of molecular carbon and a few other molecular features found in comets also occur in the visible part of the spectrum.

The gases in planetary atmospheres and in the tails and comae of comets are generally in molecular form, and therefore must be studied through their vibrational and rotational transitions, or more rarely through their electronic transitions (chapter 5). The spectroscopes used for this purpose are generally conventional instruments designed to operate at high spectral resolutions in the infrared, including echelle and Fourier transform spectroscopes (chapters 8 and 10). Long-slit or multi-object spectroscopes may also be used because the planets are resolved.

One problem with Earth-based observations is that our own atmosphere produces absorption features. These may sometimes be separated from features due to the same gases in the planetary atmosphere through Doppler shifts caused by the relative motions of the Earth and planet. But this is not always the case. The discovery that nitrogen not methane was the main constituent of Titan's atmosphere, for example, had to await the spacecraft missions to Saturn. Gases not present, or present only in low abundance in our own atmosphere, may generally be studied from the Earth relatively easily provided that their spectral features occur within the atmosphere's infrared 'windows'. In addition to nitrogen and methane, numerous other compounds of carbon and hydrogen such as C_4H_2, C_3H_4, C_2H_6 and HCN have thus been found in Titan's atmosphere, H_3^+ in Jupiter's outer atmosphere, etc.

Spacecraft-borne infrared spectroscopes such as the FTIS on Voyager (chapter 10) clearly do not suffer from the limitations imposed by the Earth's atmosphere, and so give a clearer picture of the compositions of planetary atmospheres. However, some molecules, such as H_2 and O_2, do not have rotational and vibrational spectra because they have no electric dipole moment (chapter 5), and so cannot easily be studied in the infrared (though H_2 and

HD have been detected on the Jovian planets through transitions arising via quadrupole interactions). Other constituents of atmospheres, especially the noble gases such as helium and xenon, are in atomic form. These gases have therefore to be studied from their ultraviolet spectra, and many planetary probes such as the Mariners, Pioneers and Voyagers have carried ultraviolet spectroscopes for this purpose. Planetary ultraviolet spectroscopy may also be undertaken by spacecraft in Earth orbit such as IUE and the Hubble Space Telescope (chapter 10).

15.3 PLANETARY SURFACES

The spectroscopic study of planetary surfaces in the visual and near infrared is by reflectance spectroscopy. The spectral signatures of solid materials (chapter 7) are far broader than those of molecules and atoms, and there is usually considerable ambiguity in their interpretation. Remote sensing of the surface of the Earth relies upon direct sampling (usually called obtaining ground-truth data) to calibrate the remote observations. This is clearly not possible with other planets except for those few sites on the Moon, Mars and Venus where spacecraft have landed. Interpretation of surface spectra therefore requires information from other sources to reduce the level of ambiguity. Such information may comprise surface temperatures, thermal conductivities, relative abundances of elements, surface structures, mean densities, moments of inertia, and in some cases the composition of the atmosphere. Some laboratory simulation may be possible, but often the molecules or their radicals suspected of being present on some of the planets cannot be synthesized under terrestrial conditions.

Even when some estimate of the surface composition has been achieved, this will normally only refer to the very thin top layer of the surface, and this may not be characteristic of the surface as a whole. Thin layers of frozen gases (on the satellites of the outer planets) or iron oxides (on Mars) for example, may overlie much deeper layers of silicate minerals.

Nonetheless, some results have been achieved. For example, sulphur compounds have been detected on Io, olivine, pyroxene, carbonates, etc on Mars, carbonaceous chondrite-type compositions for some asteroids, water frost on Callisto and many other satellites of the outer planets, nitrogen ice on Triton and Pluto, and methane and carbon monoxide ices on Pluto. For Io, the hot spots which are its volcanoes dominate the spectrum between about 4 and 12 μm, and can therefore be observed directly in that region.

In other regions of the spectrum, greater certainty in the results may be possible. Thus gamma ray spectroscopy and x-ray fluorescence spectroscopy have been used to study the Moon, Venus and Mars, leading to measurements of the abundances of radioactive isotopes of elements such as potassium, thorium and uranium. Microwave radiometry and radar can also be useful, especially for compounds formed from the lighter elements.

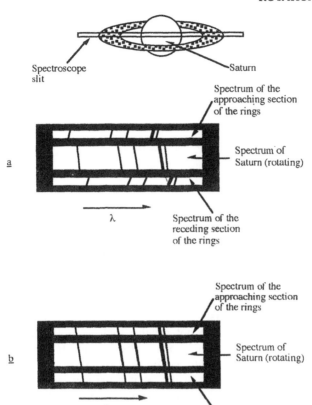

Figure 15.1 Doppler shifts in a spectrum of Saturn and its rings. (*a*) As observed, and for rings composed of individual particles orbiting the planet; (*b*) how the spectrum would appear if the rings were a solid disc.

15.4 ROTATION

Rotation of the whole planet or motions within planetary atmospheres can be studied via Doppler shifts of their spectral lines. However, the low velocities usually to be found mean that more accurate results may normally be obtained by direct observations. In the case of Saturn's rings, however, that is not possible and the Doppler shifts there may be used to show that the rings are composed of myriads of small particles orbiting the planet rather than being a solid disc (figure 15.1).

16

Nebulae and the Interstellar Medium

16.1 NEBULAE

Nebulae may be observed with conventional spectroscopes or with specially designed instruments such as slitless spectroscopes or TAURUS (chapter 10). Long-slit and multi-object spectroscopes are also useful because many nebulae are angularly resolved. Smaller and/or more distant nebulae may not be angularly resolved, and for these spectroscopy is generally essential in order to distinguish them from stars.

The nebulae divide into the hot nebulae such as H II regions, planetary nebulae, supernova remnants and nova shells, and the H I regions and cool molecular clouds. The first group may be studied at most wavelengths, but the latter are primarily the concern of the far infrared and radio astronomers. A small proportion (about 1%) of the material in the cool clouds and in some of the hot nebulae is in the form of dust particles. These are typically a few hundreds of nanometres in size and absorb extremely strongly in the visible and short wave regions of the spectrum. The properties of the H I and H II regions and cool molecular clouds are closely related to those of the interstellar medium since the former are essentially just highly concentrated portions of the latter, though strongly heated in the case of the H II regions by the embedded stars. There is also some overlap, particularly in the presence of emission lines and in the types of line profiles, with the gaseous envelopes to be found around some stars (chapter 14).

The basic optical spectrum of the hot nebulae is that of an emission line spectrum (figure 4.1). The strongest of the emission lines, however, are forbidden lines (chapters 1 and 4), such as [O II] 372.7, [Ne III] 386.9 and [O III] 436.3, 495.1 and 500.7.This is a result of the very low densities of most nebulae (typically 10^9 to 10^{12} atoms per cubic metre) and their low levels of illumination. The main excitation mechanism is therefore that of collisions between the particles. Once excited, an atom will quickly cascade back towards its ground level through allowed transitions. However, if a metastable level exists (an energy level with only forbidden transitions down to lower energy levels) then the electron may become trapped in that level. Since all excitation

242

mechanisms, including collisions, which would enable the electron to escape from the metastable level by a transition to a higher level are very rare, over time the majority of the atoms may become excited to the metastable level. The forbidden lines may thus become the strongest lines in the spectrum. Similarly in the infrared several forbidden vibrational transitions (chapter 5) from molecular hydrogen may be detected.

The allowed emission lines mostly arise from collisional ionization followed by recombination, in atoms or ions without metastable levels to trap the electrons, and are produced by hydrogen, helium, carbon, nitrogen and oxygen, etc. A few lines of doubly ionized nitrogen and oxygen arise through selective fluorescence. This latter process occurs when a line from one element coincides with a strong emission line from another element. The upper level of the first element becomes overpopulated through absorption of the emission line photons, and in turn produces emission lines as the electrons cascade back to the ground state. The ultraviolet emission lines, such as Ly $-\alpha$, are frequently the precursors in this process.

A slitless spectrum of a nebula will result in a series of monochromatic images of the nebula in each of its emission lines (figure 16.1). Differences between the images will reflect differences in excitation and ionization, or more rarely composition, across the nebula. A slit spectrogram will provide information as usual on velocities etc. If the nebula is optically thin and expanding, then the lines will be double since both the approaching and receding parts of the nebula will be seen (figure 16.2). The velocity structure of the nebula may also be studied using specialized instruments such as TAURUS (chapter 10). The expansion of a nebula may sometimes be used to estimate its distance. If the increase in the angular size of the nebula can be detected by observations over a period of years, then by assuming the expansion to be isotropic, the linear velocity may be found from the Doppler shifts of the lines in the spectrum. The distance is thereby given by

$$D = \frac{v}{5\Delta\alpha} \quad \text{pc} \qquad (16.1)$$

where v is the linear expansion velocity in $\mathrm{km\,s^{-1}}$ and $\Delta\alpha$ is the angular expansion of the nebula in arcsec/yr.

A difficulty with this method, however, is that some of the angular expansion of the nebula may not be due to the physical motion of the material, but to an ionization front moving outwards through the surrounding H I region. Distances obtained by this method can thus only be lower limits.

The physical conditions in nebulae are clearly about as far from LTE as it is possible to get. The Saha, Boltzmann, Maxwell etc equations are therefore inapplicable. Nonetheless, it is still possible to determine temperatures and densities for the material in the hot nebulae from their spectra. The temperature is found from the oxygen and nitrogen optical forbidden emission line intensities. A partial energy level diagram is shown in figure 16.3 for doubly ionized oxygen.

Nebula

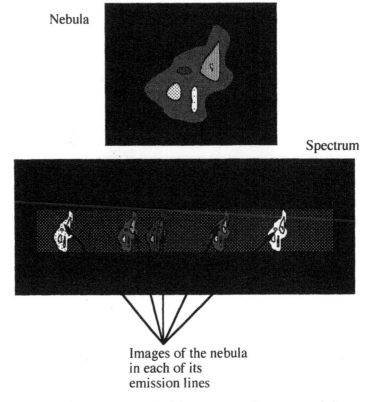

Spectrum

Images of the nebula
in each of its
emission lines

Figure 16.1 A schematic slitless spectrum of a gaseous nebula.

For densities less than about 10^{11} m^{-3}, we may neglect collisional de-excitation. Every excitation to the ^{1}S level will therefore result in the emission of either [O III] 436.3 or [O III] 232.1 photons. The relative probability of the two emissions will be given by their transition probabilities (chapter 4). Similarly, every excitation to the ^{1}D level will result in the emission of an [O III] 495.9 or [O III] 500.7 photon. The transition producing the [O III] 436.3 line also populates the ^{1}D level, but this contribution is negligible compared with the direct excitations, as may be seen from the value of the line ratios below. The excitations to both levels are by collisions and the different energies required (5.4 eV and 2.7 eV) mean that the resulting relative populations will depend upon the energies of the atoms and hence upon the kinetic temperature. The relationship between line intensities and temperature is

$$\frac{I_{495.9} + I_{500.7}}{I_{436.3}} = \frac{8.32 \exp(39\,200T^{-1})}{1 + 4.5 \times 10^{-10} N_e T^{-1/2}} \tag{16.2}$$

where N_e is the electron number density.

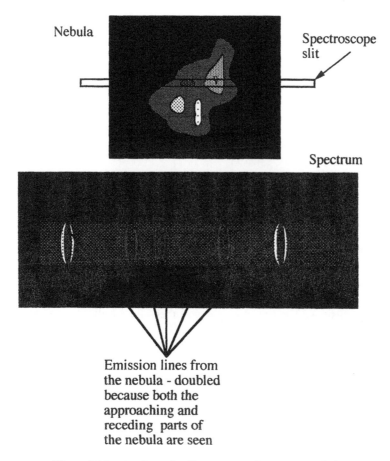

Figure 16.2 A schematic slit spectrum of a gaseous nebula.

Singly ionized nitrogen has a similar energy level structure to that of doubly ionized oxygen, and so there is a similar relationship for the forbidden nitrogen lines:

$$\frac{I_{654.8} + I_{658.3}}{I_{575.5}} = \frac{7.53 \exp(25\,000T^{-1/2})}{1 + 2.7 \times 10^{-9} N_e T^{-1/2}}. \tag{16.3}$$

This line intensity ratio for oxygen changes from a value of about 1 000 at a temperature of 7 000 K to about 100 at 15 000 K. The ratio for the nitrogen lines similarly changes from about 300 down to about 40 over the same temperature range.

Densities of the hot nebulae may also be obtained from forbidden line intensity ratios. If the upper levels of two forbidden lines have similar energies, then in the low density limit their intensity ratio will be close to unity. This is because the line strength will depend only upon the energy level populations

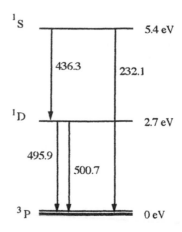

Figure 16.3 A partial energy level diagram for doubly ionized oxygen.

and these will be in the ratio of the level's statistical weights. At higher densities, the lifetimes of the levels will become longer than the mean time between collisions. The line strengths will then additionally be proportional to the transition probabilities from the levels. The ratios of the line strengths will thus vary with density. The line ratios usually used for this purpose are [O II] 372.9 / [O II] 372.6 and [S II] 671.6 / [S II] 673.1, though other lines may also be employed. The line ratios vary from about 1.5 in the low density limit to about 0.35 at high densities, but the manner of the variation is complex, and the reader is referred to more specialized books (appendix C) for further details.

Though not a part of spectroscopy itself, a consequence of the nature of the spectrum of the hot nebulae is that direct imaging of them is easier than might appear from their low overall levels of brightness. A large fraction of their energy is concentrated into a few emission lines. A narrow band filter centred on one of those emission lines will therefore greatly reduce the background noise whilst still leaving the nebula as a bright object. This is particularly useful technique for the poorer observing sites where the sky background is strongly contaminated by artificial sources.

Though not strictly part of the brief for this book, mention must be made of the 21 cm line of hydrogen because of its importance in mapping out the cool clouds and nebulae. The 21 cm line arises from a forbidden hyperfine transition of atomic hydrogen (chapter 3). The more abundant molecular hydrogen is difficult to observe directly because, lacking a dipole moment, it has no allowed rotational or vibrational transitions (chapter 5). The cool clouds in the Galaxy, and thus the shape of the Galaxy itself, have therefore been studied from observations of the atomic hydrogen line. The Doppler shifts due to the individual motions of the clouds around the Galaxy allow several clouds along the same line of sight to be distinguished. A velocity curve for the Galaxy then leads to an estimate of the cloud's distance. Since the clouds are concentrated in the spiral arms, a map of their distribution mirrors the optical appearance of the Galaxy which, of course, cannot be observed directly because of interstellar absorption (below).

16.2   THE INTERSTELLAR MEDIUM

The material between the stars is little different from that in H I regions, save that it is more rarefied. A typical density would be around 10^6 atoms per cubic metre, a factor of a thousand down on that of the gaseous nebulae. About 1% of the material is in the form of dust particles thought to have a refractory core of graphite or of silicate compounds and a mantle of frozen gases.

16.2.1   The interstellar gas

This gas is difficult to observe directly. Absorption lines produced by the interstellar gas along the line of sight may, however, be detected in the spectra of distant stars. Such lines are generally narrow resonance lines from neutral atoms and their first level of ionization. The sodium D lines and the H and K lines of ionized calcium are the strongest such interstellar lines. Numerous molecules, some quite complex, have been detected by radio and microwave spectroscopy. The majority of such molecules, however, are to be found in the cool molecular clouds, where they are protected from ionizing radiation, rather than in genuine interstellar space. Simpler molecules such as H_2, OH, CO and CH may be detected via interstellar absorption in the optical and ultraviolet spectral regions.

16.2.2   The interstellar dust

Despite its small proportion of the total interstellar material, the interstellar dust has a major effect on optical observations. This is because the dust particles are a few hundred nanometres in size and absorb optical photons extremely strongly. Our view within the plane of the Milky Way Galaxy is limited to a few thousand parsecs around the Sun by this absorption. The particles are solid objects and therefore have relatively uninformative spectra (chapter 7). The absorption is roughly proportional to $\lambda^{-4/3}$, but has features at 0.2, 3, 10 and 30 μm attributed to the composition of the particles. Broad absorption lines in the spectra of distant stars, such as that at 443 nm which is some 4 nm wide, and which are known as the Diffuse Interstellar Bands, have been attributed to the dust particles in the past, though now this seems less likely. The particles may also be observed directly via the light that they scatter, since this forms a component of the night sky background.

17

Extra-Galactic Objects

17.1 INTRODUCTION

Extra-galactic objects, and slightly paradoxically the Milky Way Galaxy is generally included within this term, emit light from their constituent stars and from the hot gas clouds within them. They may also absorb the light from more distant objects against which they may be silhouetted. Their spectroscopic observation, and much of the information that may be learnt from it, therefore closely follows that for stars (chapters 11 and 14) and nebulae (chapter 16). In particular, radial velocities and abundances are found using identical approaches to those for stars and nebulae. Some specialized techniques and methods are found, however, and all approaches must generally be modified to cope with the often low apparent brightnesses of the extra-galactic objects. High efficiency designs such as the LDSS (chapter 10) are therefore at a premium and long exposures the norm. With the best of current instrumentation and techniques, slit spectra down to a visual magnitude of +23 are just about obtainable at a reasonable dispersion for emission line objects in an exposure lasting all night. Since galaxies are often found in clusters, the various multi-object spectroscopes (chapter 10) are frequently used to obtain many spectra simultaneously. In the nearer objects, H II regions and other hot nebulae can be highlighted by observing through a narrow band filter centred on one of their emission lines.

17.2 DISTANCES

The best known result of the spectroscopy of extra-galactic objects is their redshifts and the determination of their distances from the Hubble relation. The greatest redshift currently stands at about 4.9, corresponding to a velocity (via the relativistic Doppler shift formula, equation (12.3)) of about $283\,000$ km s^{-1}, and a distance of $5\,700$ Mpc for a Hubble constant of 50 km s^{-1} Mpc^{-1}.

Rather less well known is the Tully–Fisher relation, which determines the distance of a spiral galaxy from an empirical relationship between its absolute magnitude and the maximum rotational velocity:

248

$$M \approx -5 \log_{10} V_M - 8 \qquad (17.1)$$

where M is the visible integrated absolute magnitude and V_M is the maximum rotational velocity in $km\,s^{-1}$. The distance is obtained from the absolute magnitude formula (equation (14.1)) once the visible integrated apparent magnitude of the galaxy has been measured. For more distant galaxies, the rotational velocity is determined from the width of the 21 cm line rather than by direct measurement.

17.3 SPECTRA

The majority of extra-galactic objects are moving away from us. As mentioned above this leads to one of the main methods of estimating their distances. It also, however, means that the observed spectra in the visual region originated at shorter wavelengths. For example, at a redshift of 3, the observed wavelengths over the region 400–700 nm would have been emitted between 100 and 175 nm. For many extra-galactic objects therefore, we may observe their ultraviolet and far ultraviolet spectra via optical spectroscopy. The Lyman-α line is often strongly in emission, especially for active galaxies, quasars and similar objects. An efficient search for such objects may thus be made by using an objective prism on a Schmidt camera. There will be many thousands of spectra on such a plate, but the distant quasars etc will be easily picked out from the rest through the presence of the strong Ly-α emission (figure 17.1). Cool gas clouds and galactic haloes along the line of sight to a distant object will result in many narrow absorption lines at lesser redshifts, also due to Ly-α. These may sometimes be so numerous that they are referred to as the Ly-α forest.

For the nearer objects, we observe most or all of the normally visible spectrum. Then, with many galaxies, the spectrum is just the aggregate sum of the spectra of their constituent stars. Within the spectra of active galaxies, however, emission lines are frequently to be found. Often these are due to forbidden transitions. In quasars and Seyfert 1 galaxies, the allowed emission lines have velocity widths of $10\,000\ km\,s^{-1}$ or so. The forbidden lines have widths of about $500\ km\,s^{-1}$. Many of the lines arise from highly ionized states such as Fe VII to Fe XIV, Si VII and Ca VIII. Such lines are often called Coronal because of their appearance in the solar corona (chapter 14). In Seyfert 2 galaxies, all the lines are around $500\ km\,s^{-1}$ in width, while in LINERS (Low Ionization Nuclear Emission Regions), the levels of ionization are much lower, as the acronym suggests.

The active galaxies are often very strong emitters in the short wave and radio regions of the spectrum, with synchrotron radiation dominating in the latter. The overall spectrum of active galaxies is roughly constant in terms of energy per frequency decade, from the radio region to gamma rays. The nuclei of more normal galaxies, including our own, may have similar emissions on a

Stellar spectra   Quasar's
                                   spectrum

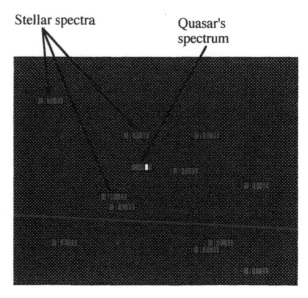

Figure 17.1   Schematic objective prism field containing a quasar.

smaller scale. The outer regions of galaxies generally have a roughly black-body spectrum, peaking in the visible or near ultraviolet. Forbidden iron lines are an indicator of a starburst galaxy, and probably arise from the numerous supernovae occurring within them.

17.4   AGES

The stars in a globular cluster will all have formed within a short time span. An H/R diagram (chapter 11, figure 14.1) of a single cluster will thus only contain stars on its main sequence up to the point at which the main sequence lifetime equals the age of the cluster. More massive stars will have evolved away towards the giant region (figure 17.2). The spectral type at this cut-off point on the main sequence can then be used to determine the age of the globular cluster (table 17.1).

17.5   SIZES

The emission lines of many active galaxies, including quasars, Seyferts, BL Lacs etc, vary on time scales of years or less, and in some cases the total luminosity may change as well. This limits the sizes of the emission regions, because in general an object cannot vary in less than the time it takes for light to cross it. There are some minor exceptions to this rule, such as a thin sheet of material

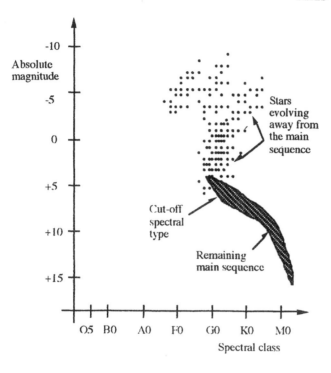

Figure 17.2 Schematic H/R diagram for a globular cluster.

Table 17.1 Main sequence lifetimes.

Spectral class of cut-off point	Lifetime (years)
B0	2×10^6
A0	2×10^8
F0	2×10^9
G0	8×10^9
K0	3×10^{10}

perpendicular to the line of sight and illuminated by another object, but these are unimportant when we are considering many similar examples, because of the tight geometrical constraints on such exceptions if they are to work. The emission regions of active galaxies, which may have luminosities up to 10^{14} times that of the Sun (10 000 times the luminosity of the Milky Way Galaxy), are thus limited to a few light-years, and in some cases to a few light-weeks, in size.

17.6 BACKGROUND RADIATION

Though not strictly part of the brief of this book, mention must finally be made of the most distant extra-galactic 'object' yet observed spectroscopically, the Microwave Background Radiation. This radiation, which has a black-body distribution peaking at 2.7 K, pervades the universe and is the remnant of the Big Bang fireball. In a sense it can be observed by almost anyone, since between 0.1% and 1% of the noise from a domestic television, de-tuned from any station, will be due to background photons. The radiation originated about 300 000 years after the Big Bang at a temperature of about 3000 K, and therefore has a redshift of about 1000!

Appendix A

Conversion Formulae

The following conversion formulae may be found useful. The notation $E(J)$ denotes energy measured in joules etc.

Energy

$$E(J) = 1.98649 \times 10^{-23} E(cm^{-1}) = 1.602192 \times 10^{-19} E(eV)$$

$$E(eV) = 6.24145 \times 10^{18} E(J) = 1.23985 \times 10^{-4} E(cm^{-1})$$

$$E(cm^{-1}) = 8.06549 \times 10^{3} E(eV) = 5.03401 \times 10^{22} E(J)$$

Frequency

$$\nu(Hz) = 2.99793 \times 10^{10} E(cm^{-1})$$
$$= 2.41798 \times 10^{14} E(eV)$$
$$= 1.50916 \times 10^{33} E(J)$$

Wavelength

$$\lambda(nm) = 1.98649 \times 10^{-16} / E(J)$$
$$= 10^{7} / E(cm^{-1})$$
$$= 1.23985 \times 10^{3} / E(eV)$$

Differential formulae

$$\Delta\lambda(nm) = -1.98649 \times 10^{-16} E(J)^{-2} \Delta E(J)$$
$$= -10^{7} E(cm^{-1})^{-2} \Delta E(cm^{-1})$$
$$= -1.23986 \times 10^{3} E(eV)^{-2} \Delta E(eV)$$
$$= -2.99793 \times 10^{8} \nu(Hz)^{-2} \Delta\nu(Hz)$$

Appendix B

Term Formation of Equivalent Electrons

The argument for the formation of terms for two equivalent p electrons ($l = 1$) is given in the table below.

Table B.1

	m_{l1}	m_{l2}	m_{s1}	m_{s2}	
1	+1	+1	+1/2	+1/2	Not allowed: all four quantum numbers identical
2	+1	+1	+1/2	−1/2	Allowed: contributes to the $^1D_2$ term ($L = 2$, $S = 0$)
3	+1	+1	−1/2	+1/2	Not allowed: same configuration as line 2
4	+1	+1	−1/2	−1/2	Not allowed: all four quantum numbers identical
5	+1	0	+1/2	+1/2	Allowed: contributes to the $^3P_2$ level ($L = 1$, $S = 1$)
6	+1	0	+1/2	−1/2	Allowed: contributes to the $^1D_2$ term ($L = 2$, $S = 0$)
7	+1	0	−1/2	+1/2	Allowed: contributes to the $^3P_1$ level ($L = 1$, $S = 1$)
8	+1	0	−1/2	−1/2	Allowed: contributes to the $^3P_0$ level ($L = 1$, $S = 1$)
9	+1	−1	+1/2	+1/2	Allowed: contributes to the $^3P_2$ level ($L = 1$, $S = 1$)
10	+1	−1	+1/2	−1/2	Allowed: contributes to the $^3P_1$ level ($L = 1$, $S = 1$)
11	+1	−1	−1/2	+1/2	Allowed: contributes to the $^1S_0$ term ($L = 0$, $S = 0$)
12	+1	−1	−1/2	−1/2	Allowed: contributes to the $^3P_0$ level ($L = 1$, $S = 1$)
13	0	+1	+1/2	+1/2	Not allowed: same configuration as line 5
14	0	+1	+1/2	−1/2	Not allowed: same configuration as line 6

Table continued on next page

Table B.1 *(continued.)*

	m_{l1}	m_{l2}	m_{s1}	m_{s2}	
15	0	+1	−1/2	+1/2	Not allowed: same configuration as line 7
16	0	+1	−1/2	−1/2	Not allowed: same configuration as line 8
17	0	0	+1/2	+1/2	Not allowed: all four quantum numbers identical
18	0	0	+1/2	−1/2	Allowed: contributes to the $^1D_2$ term $(L = 2, S = 0)$
19	0	0	−1/2	+1/2	Not allowed: same configuration as line 18
20	0	0	−1/2	−1/2	Not allowed: all four quantum numbers identical
21	0	−1	+1/2	+1/2	Allowed: contributes to the $^3P_2$ level $(L = 1, S = 1)$
22	0	−1	+1/2	−1/2	Allowed: contributes to the $^1D_2$ term $(L = 2, S = 0)$
23	0	−1	−1/2	+1/2	Allowed: contributes to the $^3P_1$ level $(L = 1, S = 1)$
24	0	−1	−1/2	−1/2	Allowed: contributes to the $^3P_0$ level $(L = 1, S = 1)$
25	−1	+1	+1/2	+1/2	Not allowed: same configuration as line 9
26	−1	+1	+1/2	−1/2	Not allowed: same configuration as line 10
27	−1	+1	−1/2	+1/2	Not allowed: same configuration as line 11
28	−1	+1	−1/2	−1/2	Not allowed: same configuration as line 12
29	−1	0	+1/2	+1/2	Not allowed: same configuration as line 21
30	−1	0	+1/2	−1/2	Not allowed: same configuration as line 22
31	−1	0	−1/2	+1/2	Not allowed: same configuration as line 23
32	−1	0	−1/2	−1/2	Not allowed: same configuration as line 24
33	−1	−1	+1/2	+1/2	Not allowed: all four quantum numbers identical
34	−1	−1	+1/2	−1/2	Allowed: contributes to the $^1D_2$ term $(L = 2, S = 0)$
35	−1	−1	−1/2	+1/2	Not allowed: same configuration as line 34
36	−1	−1	−1/2	−1/2	Not allowed: all four quantum numbers identical

Thus the allowed terms are $^1D_2$, $^3P_{2,1,0}$ and $^1S_0$.

Appendix C

Bibliography

A selection of books for further reading is given below. This is not intended to be a complete list but just to provide the interested reader with a starting point for further work.

GENERAL

Astrophysical Quantities C W Allen, Athlone Press, Third Edition 1973.
A Dictionary of Spectroscopy R C Denney, John Wiley & Sons, 1982.
Glossary of Astronomy and Astrophysics J Hopkins, University of Chicago Press, 1980.
Astrophysical Formulae Second Edition K R Lang, Springer-Verlag, 1980.
Telescopes and Techniques C Kitchin, Springer-Verlag, 1995.

HISTORY OF SPECTROSCOPY

The Analysis of Starlight J B Hearnshaw, Cambridge University Press, 1990.
Their Majesties' Astronomers C A Ronan, Bodley Head, 1967.

QUANTUM MECHANICS

The Quantum Theory of Light R Loudon, Clarendon Press, 1986.
Introduction to Quantum Mechanics P T Matthews, McGraw-Hill, 1974.
Quantum Mechanics; An Introduction W Greiner, Springer-Verlag, 1994.
Introduction to Quantum Mechanics D Griffiths, Prentice Hall, 1994.

EQUIPMENT AND TECHNIQUES

Astrophysical Techniques C Kitchin, Adam Hilger, 1991.
Fibre Optics in Astronomy S C Barden, Astronomical Society of the Pacific, 1988.

Radio Astronomy J D Kraus, Cygnus-Quasar Books, 1986.

Photodetectors P Dennis, Plenum Press, 1985.

Deconvolution of Absorption Spectra W E Blass and G W Halsey, Academic Press, 1981.

Inverse Problems in Astronomy I J D Craig and J C Brown, Adam Hilger, 1986.

Observational Astronomy for Amateurs ed P Moore, Springer-Verlag, 1994.

Getting the Measure of Stars W Cooper and E Walker, Adam Hilger, 1980.

The Astronomy Source Book: The Complete Guide to Astronomical Equipment, Publications, Planetariums, Organisations, Events and More R Gibson, Woodbine House, 1992.

REMOTE SENSING

Introduction to Remote Sensing J B Campbell, Guildford Press, 1987.

Introduction to Environmental Remote Sensing E Barrett and L Curtis, Chapman & Hall, 1992.

Introduction to Remote Sensing A Cracknel and L Hayes, Taylor & Francis, 1991.

ASTROPHYSICS

Introduction to Stellar Atmospheres and Interiors E Novotny, Oxford University Press, 1973.

Stellar Atmospheres D Mihalas and W H Freeman, 1978.

Astrophysics; The Atmospheres of the Sun and Stars L H Aller, Ronald Press, 1963.

Astrophysics: Nuclear Transformations, Stellar Interiors, and Nebulae L H Aller, Ronald Press, 1954.

Astrophysics 1: Stars R Bowers and T Deeming, Jones & Bartlett, 1984.

Astrophysics 2: Interstellar Matter & Galaxies R Bowers and T Deeming, Jones & Bartlett, 1984.

Chemistry of the Solar System H E Suess, John Wiley & Sons, 1987.

Early Emission Line Stars C Kitchin, Adam Hilger, 1982.

Stars, Nebulae and the Interstellar Medium C Kitchin, Adam Hilger, 1987.

The HR Diagram A G Davis Phillip and D S Hayes, D Reidel, 1978.

Chemistry of Atmospheres R P Wayne, Clarendon Press, 1985.

Radiative Processes in Astrophysics G B Rybicki and A P Lightman, John Wiley & Sons, 1979.

Radiation Processes in Astrophysics W H Tucker, MIT Press, 1975.

High Energy Astrophysics Volume 1: Particles. Photons and their Detectors 1992, *Volume 2: Stars, the Galaxy and the Interstellar Medium* M S Longair, Cambridge University Press, 1994.

Discovering the Secrets of the Sun R Kippenhahn, John Wiley & Sons, 1994.

Physics of Stars A Phillips, John Wiley & Sons, 1994.

ASTRONOMICAL SPECTROSCOPY

Astronomical Spectroscopy A D Thackeray, Eyre & Spottiswood, 1961.

Interpretation of Spectra and Atmospheric Structure in Cool Stars Y Fujita, University of Tokyo Press, 1970.

Spectral Classification and Multicolour Photometry C Fehrenbach and B E Westerlund, D Reidel, 1973.

The Impact of Very High Signal to Noise Spectroscopy on Stellar Physics ed G Cayrel de Strobel and M Spite, Kluwer, 1988.

Index of Galaxy Spectra G Gisler and E Friel, Parchart, 1979.

Stellar Spectroscopy: Peculiar Stars M Hack and O Struve, Osservatorio Astronomico di Trieste, 1970.

Stars and their Spectra J B Kaler, Cambridge University Press, 1989.

Catalogue of Emission Lines in Astrophysical Objects A B Meinel, A F Aveni and M W Stockton, University of Arizona Press, 1969.

MK Spectral Classifications: Seventh General Catalogue W Buscombe, Northwestern University Press, 1988.

The Classification of Stars C Jaschek and M Jaschek, Cambridge University Press, 1990.

Atlas of Representative Stellar Spectra Y Yamashita, K Nariai and Y Norimoto, University of Tokyo Press, 1977.

UV and X-ray Spectroscopy of Laboratory and Astrophysical Plasmas E Silver and S Kahn, Cambridge University Press, 1993.

Atoms, Ions and Molecules: New Results in Spectral Line Astrophysics ed A Haschick and P Ho, Astronomical Society of the Pacific, 1991.

Spectroscopy of Astrophysical Plasmas ed A Dalgarno, D Layzer, Cambridge University Press, 1987.

Detection and Spectrometry of Faint Light J Meaburn, Kluwer, 1980.

Astronomical Infrared Spectroscopy ed S Kwok, Astronomical Society of the Pacific, 1993.

Astrophysical and Laboratory Spectroscopy R Brown and J Lang, Institute of Physics, 1987.

Atmospheric Spectroscopy ed G Hunt and J Ballard, Elsevier, 1985.

ATOMIC SPECTROSCOPY

Structure and Spectra of Atoms W G Richards and P R Scott, John Wiley & Sons, 1976.

An Introduction to Atomic Absorption Spectroscopy L Ebdon, Heyden, 1982.

Basic Principles of Spectroscopy R Chang, McGraw-Hill, 1971.

Spectroscopy volumes 1, 2, and 3 B P Straughan and S Walker, Chapman Hall, 1976.

Spectral Line Formation J T Jefferies, Blaisdell, 1968.

The Theory of Atomic Spectra E U Condon and G H Shortley, Cambridge University Press, 1967.

Atomic Absorption Spectroscopy J W Robinson, Edward Arnold, 1966.

Atomic Spectra and Atomic Structure G Herzberg, Dover, 1944.

Introduction to Atomic Spectra H E White, McGraw-Hill, 1934.

Visual Methods of Emission Spectroscopy N Sventitskii, Coronet, 1965.

Introduction to the Theory of X-ray and Electronic Spectra of Free Atoms R Karazija, Plenum 1992.

Introduction to the Spectroscopy of Atoms P Heckman and E Trabert, Elsevier, 1989.

MOLECULAR SPECTROSCOPY

Structure and Spectra of Molecules W G Richards and P R Scott, John Wiley & Sons, 1985.

Introduction to Molecular Spectroscopy G M Barrow, McGraw-Hill, 1962.

Spectrophysics A P Thorne, Chapman & Hall, 1974.

Molecular Spectra and Molecular Structure: 1 Spectra of Diatomic Molecules G Herzberg, Van Nostrand, 1963.

Introduction to Molecular Spectroscopy E F H Brittain, W O George and C H J Wells, Academic Press, 1970.

Molecular Spectroscopy I N Levine, John Wiley & Sons, 1975.

Vibrational Spectroscopy A Fadini and F-M Schnepel, Ellis Horwood, 1989.

The Spectra and Structure of Simple Free Radicals: An Introduction to Molecular Spectroscopy G Herzberg, Dover, 1988.

Molecules and Radiation: An Introduction to Modern Molecular Spectroscopy J Steinfeld, MIT Press, 1978.

Introduction to Molecular Spectroscopy G Barrow, McGraw-Hill, 1991.

ATOMIC ENERGY LEVELS AND TRANSITION PROBABILITIES

Atomic Energy Levels volumes 1, 2 and 3 C E Moore, US National Bureau of Standards, NSRDS-NBS 35, 1971.

A Multiplet Table of Astrophysical Interest C E Moore, NBS Technical Note 36, 1959.

An Ultra-violet Multiplet Table C E Moore, NBS Circular 488, 1950.

Lines of the Chemical Elements in Astronomical Spectra P W Merrill. Carnegie Institute of Washington Publication 610, 1958.

Bibliography on Atomic Transition Probabilities (1914 through October 1977) J Fuhr, B Miller and G Martin, NBS Special Publication 505, 1978, and Supplement 1, NBS Special Publication SP 505-1, 1980.

Bibliography on Atomic Energy Levels and Spectra NBS Special Publication 363, 1980, and Supplements 1, 2 and 3, NBS Special Publications SP 363 S1 1971, SP 363 S2 1980, SP 363 S2 1985.

Tables of Spectral Lines of Neutral and Ionised Atoms A R Striganov and N S Sventitskii, Plenum Press, 1968.

Wavelengths and Transition Probabilities for Atoms and Atomic Ions J Reader, C H Corliss, W L Wiese and G A Martin, NBS Monograph 68, 1980.

Experimental Transition Probabilities for Spectral Lines of Seventy Elements C Corliss and W Bozmann, NBS Monograph 53, 1962.

Appendix D

Constants

The values of some of the commonly used constants referred to in this book are tabulated below, for ease of reference.

Table D.1

Constant	Symbol	Value	First occurrence
Bohr radius of hydrogen	a_0	5.295×10^{-11} m	Equation (2.20)
Boltzmann's constant	k	1.381×10^{-23} J K^{-1}	Equation (4.5)
Charge on the electron	e	1.602×10^{-19} C	Equation (2.2)
Electron-volt	eV	1.602×10^{-19} J	Table 2.3
Fine structure constant	α	7.297×10^{-3}	Equation (2.18)
Larmor frequency	ν_L	1.400×10^{10} Hz	Chapter 2
Mass of the electron	m_e	9.110×10^{-31} kg	Equation (2.1)
Permittivity of free space	ε_0	8.854×10^{-12} F m^{-1}	Equation (2.2)
Planck constant	h	6.626×10^{-34} J s	Equation (1.4)
Rydberg constant	R	3.288×10^{15} Hz	Equation (1.3)
Velocity of light in vacuo	c	2.998×10^8 m s^{-1}	Equation (1.4)

Index